Green Energy and Technology

More information about this series at http://www.springer.com/series/8059

Dario Marra · Cesare Pianese
Pierpaolo Polverino · Marco Sorrentino

Models for Solid Oxide Fuel Cell Systems

Exploitation of Models Hierarchy
for Industrial Design of Control
and Diagnosis Strategies

 Springer

Dario Marra
Department of Industrial Engineering
University of Salerno
Fisciano, Salerno
Italy

Pierpaolo Polverino
Department of Industrial Engineering
University of Salerno
Fisciano, Salerno
Italy

Cesare Pianese
Department of Industrial Engineering
University of Salerno
Fisciano, Salerno
Italy

Marco Sorrentino
Department of Industrial Engineering
University of Salerno
Fisciano, Salerno
Italy

ISSN 1865-3529
Green Energy and Technology
ISBN 978-1-4471-7382-3
DOI 10.1007/978-1-4471-5658-1

ISSN 1865-3537 (electronic)

ISBN 978-1-4471-5658-1 (eBook)

Contents

Nomenclature

Acronyms

AC	Alternating Current
AI	Artificial Intelligence
APU	Auxiliary Power Unit
AS	Anode-Supported
BoP	Balance of Plant
CHP	Combined Heat and Power
CPO	Catalytic Partial Oxidation
CPU	Central Processing Unit
CS	Cold-Start
CS2WU	Cold-Start to Warmed-Up
DC	Direct Current
DIR	Direct Internal Reforming
DOD	Depth of Discharge
DoE	Design of Experiment
EIS	Electrochemical Impedance Spectroscopy
ES	Electrolyte-Supported
EU	European Union
FC	Fuel Cell
FCH-JU	Fuel Cells and Hydrogen Joint Undertaking
FCS	Fuel Cell System
FDI	Fault Detection and Isolation
FSM	Fault Signature Matrix
FTA	Fault Tree Analysis
GT	Gas Turbine
ICE	Internal Combustion Engine
IEA	International Energy Agency
ISM	Integrated Stack Module
LS	Least Squares

LSM	Strontium-Doped Lanthanum Manganite
MIMO	Multi-Input Multi-Output
MLPFF	Multi Layer Perceptron Feed Forward
MLR	Multi Linear Regression
MSE	Mean Squared Error
NN	Neural Network
ODE	Ordinary Differential Equation
PEM	Proton Exchange Membrane
PI	Proportional Integral
PID	Proportional Integral Derivative
PSO	Particle Swarm Optimization
RBF	Radial Basis Function
RC	Resistor-Capacitor
REF	Prereformer Conversion Factor
RNN	Recurrent Neural Network
RUL	Remaining Useful Life
SI	Splitting Index
SOC	State of Charge
SOFC	Solid Oxide Fuel Cell
SVM	Support Vector Machine
TES	Thermal Storage System
WU	Warmed-Up
YSZ	Yttria Stabilized Zirconia

Roman Symbols

A	Area (m^2)
A_s	Heat Transfer Area (m^2)
ASR	Area Specific Resistance (A cm^2)
AU	Air Utilization (–)
c	Specific Heat Capacity (J kg^{-1} K^{-1})
C	Heat Capacity (J K^{-1})
C_c	Heat Capacity of Cold Fluid (J K^{-1})
$\dot{C}_c$	Thermal Mass Flow of Cold Fluid (W K^{-1})
$\dot{C}_f$	Thermal Mass Flow of Hot Fluid (W K^{-1})
C_h	Heat Capacity of Hot Fluid (J K^{-1})
c_p	Specific Heat Capacity at Constant Pressure (J kg^{-1} K^{-1})
D_h	Equivalent Diameter (m)
$\dot{E}$	Enthalpic Power Flow (W)
$\dot{E}_{el}$	Electrical Power Flow (W)
E_{Nernst}	Nernst Ideal Potential (V)
F	Faraday Constant (C mol^{-1})

G	Gibbs Free Energy (J mol^{-1})
h	Specific Enthalpy (J mol^{-1})
$\bar{h}$	Convective Heat Transfer Coefficient (W m^{-2} K^{-1})
H	Heat Convective Coefficient (W m^{-2} K^{-1})
h_{ch}	Channel High (m)
$\bar{h}_f^0$	Specific Enthalpy of Formation (J mol^{-1})
HHV	Higher Heating Value (J kg^{-1})
I	Current (A)
J	Current Density (A cm^{-2})
$\bar{J}$	Average Current Density (A cm^{-2})
J_0	Exchange Current Density (A cm^{-2})
J_{as}	Anode Limit Current Density (A cm^{-2})
J_{cs}	Cathode Limit Current Density (A cm^{-2})
k	Thermal Conductivity (W m^{-1} K^{-1})
l	Length (m)
LHV	Lower Heating Value (J kg^{-1})
m	Mass (kg)
$\dot{m}$	Mass Flow (kg s^{-1})
$\dot{n}$	Molar Flow (mol s^{-1})
N	Computational Elements (–)
n_e	Number of electrons (–)
N_u	Nusselt Number (–)
p	Pressure (Pa)
P	Power (W)
P_{batt}	Battery Power (W)
P_{cp}	Compressor Power (W)
P_{gross}	Gross Power (W)
P_{heat}	Heat Power (W)
$\bar{P}_{heat,dwell}$	Average Heat Power Demand (W)
P_{load}	Power Demand (W)
P_{net}	Net Power (W)
$\dot{Q}$	Heat Flow (W)
$\dot{r}$	Reaction Rate (mol s^{-1})
R	Universal Gas Constant (J m^{-1} K^{-1})
R_{in}	Battery Internal Resistance (Ω)
t	Time (s)
T	Temperature (K)
U_f	Fuel Utilization (–)
V	Voltage (V)
V_0	Battery Open Circuit Voltage (V)
w_{ch}	Channel Width (m)
$\dot{W}$	Mechanical Power (W)
x	Molar Fraction (%)

Greek Symbols

α	Charge Transfer Coefficients (–)
β	Compressor Ratio (–)
Δ	Change
η	Efficiency (–)
λ	Excess of Air (–)
μ	Micro
ξ	Fault Magnitude Coefficient
ρ	Mass Density (kg m^{-3})
σ	Ionic/Electronic Conductivity (S cm^{-1})
τ	Relaxation Time (s)
Ω	Control Volume (m^3)

Footers

a	Air
Act	Activation
an	Anode
aph	Air Preheater
ca	Cathode
cer	Ceramic
ch	Channel
cm	Compressor Motor
Conc	Concentration
cond	Conductive
conv	Convective
cp	Compressor
eff	Effective
el	Electrolyte
eq	Equivalent
ext	External
f	Fuel
front	Frontal
furnace	Furnace
HE	Heat Exchanger
in	Inlet
int	Interconnect
max	Maximum
min	Minimum
Ohm	Ohmic
out	Outlet
ox	Oxidation reaction
Pb	Postburner

pre	Prereformer
pre	Preheater
prod	Product
react	Reactant
ref	Reforming reaction
s	Solid
shift	Water-gas shift reaction
stack	Stack

List of Figures

List of Tables

Chapter 1
Introduction

1.1 Solid Oxide Fuel Cells

Among nonconventional energy conversion systems, fuel cells are very promising and many applications are running today to assess both proton exchange membrane- and solid oxide-based technologies, which are referred as PEM and SOFC, respectively. Other types of fuel cells (FCs) are also available, however, their maturity is not appealing yet because of either low performance or limited industrialization potentialities. Only direct methanol fuel cells have shown some possibilities as small generator for portable devices and the setting up of a niche market has envisaged. However, despite the advantage of using liquid fuel, which is easily transportable, some problems (e.g., low efficiency and fast degradation) make these equipments scarcely appealing for large industrialization. Therefore, further studies are needed to get these devices to the maturity level comparable to that exhibit by PEM and SOFCs.

Figure 1.1 describes the basic working principle of an SOFC fed with pure H_2. A single cell consists of three main components: an anode, a cathode, and a solid electrolyte separating the two electrodes. Oxygen and hydrogen (i.e., the reactants) are supplied to cathode and anode, respectively. Under electrical load, at the cathode surface the presence of perovskite catalyst enables oxygen ionization. The solid electrolyte permits the flux of oxygen ions to the anode, where they electrooxidize hydrogen, thus releasing heat, water, and electrons. Since electrolyte material ensures quasi-zero electronic conductivity, electrons are forced to flow through interconnect and external load toward the cathode, thus closing the electrical loop.

Due to the high working temperatures, SOFCs were demonstrated to be suitable for internally reforming hydrogen-rich fuels (i.e., methane). Therefore, in case of such a fuel feed, at the interface between anode and electrolyte further reactions take place, namely methane reforming and water-gas shift reaction. Moreover, also

© Springer-Verlag London 2016
D. Marra et al., *Models for Solid Oxide Fuel Cell Systems*,
Green Energy and Technology, DOI 10.1007/978-1-4471-5658-1_1

Fig. 1.1 Schematic of SOFC
working principle

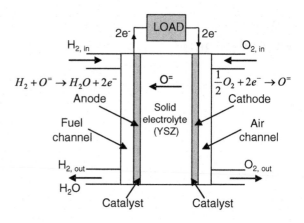

$$H_2 + O^= \rightarrow H_2O + 2e^-$$

electrooxidation of CO occurs, although with a negligible effect (Vandersteen et al. 2004), thus contributing to electric power generation (Ormerod 2003).

1.1.1 Cell Materials

As it emerged from working principle description, specific materials must be selected for electrolyte, electrodes and interconnect. Basically, due to the high working temperatures (i.e., in the range 650–1000 °C), all components are made of ceramic-based materials. The most critical component is the electrolyte, which has to guarantee, at the same time, high ionic conductivity and quasi-zero electronic conductivity. Nowadays, the most relevant material for SOFC electrolyte is Yttria Stabilized Zirconia (YSZ). Such a compound is obtained by doping zirconia with yttria at very high temperatures (i.e., 2400 °C) (Ormerod 2003). Through this process, a considerable number of oxygen ion vacancies are introduced in the original crystalline structure, thus significantly increasing the ionic conductivity at high temperatures (Bogicevic et al. 2001). Larminie and Dicks (2003) indicated that a minimum value of 10^{-2} S cm^{-1} must be guaranteed to attain acceptable power densities. Therefore, YSZ electrolyte SOFC has to be operated at temperature at least as high as 700–750 °C (Weber and Tiffeè 2004).

Anode in SOFCs must comply with conflicting requirements, such as high electronic conductivity, resistance to both reducing and oxidizing species, good thermal match with electrolyte and high porosity. These characteristics are achieved with nickel cermet anodes, obtained by adding nickel particles to YSZ. Thermal mismatch of components is a major concern in SOFC, because the risk of mismatch increases with amount of nickel added. Therefore, a satisfactory compromise between the above mentioned features has to be found. Thermal mismatch is also one of the main motivations that push SOFC researchers to develop appropriate dynamic simulation tools (Selimovic et al. 2005; Sorrentino et al. 2008; McLarty et al. 2013). During transients, the intrinsic mismatch level may cause components crack if temperature rise across the cell is not adequately controlled. Another issue to be accounted for, when selecting anode material, is the low resistance of nickel, a noble metal, to carbon deposition. This is particularly a concern in case of internal reforming of methane. Researchers are currently working either on adding further dopants or introducing a separate catalyst for the activation of the methane reforming reaction (Ormerod 2003).

Regarding cathode materials, they are required to guarantee the same characteristics as anode ones, but in addition they must promote the formation of oxygen ions at the interface with the electrolyte. To this end, the majority of SOFC developers utilize a perovskite material, Strontium-doped Lanthanum Manganite (LSM). Similar considerations on thermal mismatch mentioned for anode must be taken in account for cathode as well (Ormerod 2003).

Finally, the primarily requirement for interconnect is the high electronic conductivity. Considering the operating temperatures, ceramic materials such as lanthanum chromite are mainly used (Ormerod 2003). In some applications (Kneidel et al. 2004) multilayer ceramic interconnects with current carrier vias were successfully adopted. For relatively low operating temperatures, such as in anode-supported SOFCs, metallic interconnects also are suitable (Christiansen et al. 2005).

1.1.2 Cell Designs

Two main SOFC designs are currently underdevelopment, tubular, and planar. After an initial strong interest given to tubular designs, developers are currently focusing on planar designs. The reason for such a choice is that planar designs have lower electrical resistance than tubular. On the other hand, the tubular design takes advantage of being a more established technology, thus guaranteeing lower performance degradation during lifetime. Nevertheless, recent studies reported the fast improvements achieved for planar SOFCs on performance degradation (Blum et al. 2005; Singhal 2002).

For planar SOFCs, co-, counter- and cross-flow configurations are used. Among them, co-flow is receiving major attention, thanks to the easy monitoring of temperature gradients and the absence of dangerous hot spots (Aguiar et al. 2004).

As it is well known (Singhal 2002), lowering operating temperatures might provide numerous enhancements toward SOFC diffusion. Two strategies are currently under investigation to achieve this target. The first is the testing of new electrolyte materials, such as gadolinia-doped ceria (i.e., $Ce0.9Gd0.2O1.95$), that are capable of providing the required ionic conductivity at lower temperatures (i.e., 500 °C), as compared to YSZ electrolytes (Ormerod 2003). The second strategy, currently the most widely adopted, is the use of Anode-Supported (AS) cells. Such a design is characterized by thinner electrolytes than Electrolyte Supported (ES).

AS allows significantly reducing Ohmic losses, thus enabling low temperatures operation while guaranteeing satisfactory power density, reaching values as high as 1 W cm^{-2} (Cho et al. 2011).

1.2   Toward Industrial Deployment of Reliable SOFC Systems

Considerable technical and economical efforts have been spent from the beginning of the new century to improve the SOFCs performance giving remarkable chances as stationary energy generators in cogeneration applications (Napoli et al. 2015). Mobile uses are also potentially feasible as either auxiliary power unit (APU) (DESTA 2010; FCGEN 2011; SAFARI 2014) or range extender for the battery recharging in hybrid vehicles (Aguiar et al. 2007; Pollet et al. 2012). In the long term scenario SOFC applications could be reasonably extended to marine, rail and airplane APUs, high-power stationary generators and even to marine and rail propulsion. Although PEMFCs are more suitable when transient load operations are concerned, SOFCs give a great opportunity for both stationary and mobile uses because of their "fuel flexibility" thanks to the broad variety of chemical energy carriers that may be fed to the stack: from pure hydrogen to partial reformate gases, produced either from liquid (e.g., propane, methanol, diesel, gasoline) or natural gas (Barelli et al. 2011). This latter feature is an added value that can give more opportunities to deploy the SOFCs on markets other than those considered so far, i.e., medium/low power cogeneration. Moreover, other key characteristics of SOFCs are: high electrical efficiency, high temperature exhausted heat, modularity and no noise. On the other hand harmful emissions are not considered as an issue and the CO_2 released by the hydrogen production process is the main concern; however some NO_x, CO, and particulate might be released when residual fuel is burned. It is however worth to remember that, as far as the pollutant emissions concern, all FC technologies are the cleanest among those that use the chemical energy as input.

The big challenges to foster the diffusion of SOFC systems are mainly related to costs, durability, and performance. Focusing on cogeneration applications, the most reliable data give cost variability in the range 4500–8000 Euro/kW whereas durability spans from 10000 to 40000 h (Blum et al. 2013). To make the SOFC technology ready for the market the costs should be cut of one third, whereas durability has been targeted to 60000 h for 2020 (Kadovaki 2015; Papageorgopulos 2015). Also the efficiency of the fuel cell system is seen as a key factor and an increase of 5–10 % is considered feasible toward 2020; this improvement would bring the efficiency of CHP applications up to 90 %. The combination of the above enhancements is foreseen by the Joint Programme on Fuel Cells and Hydrogen of the European Energy Research Alliance (EERA) and could make the unit cost of the electric energy comparable to that of the grid, achieving thus the so-called "grid parity."

As far as the cost concerns a reduction may be achieved through mass production technologies, whose capital costs investment could be sustained by a larger market. Toward this objective, several public funded projects are ongoing in Europe, Japan, South Korea, and North America (Kadowaki 2015; White et al. 2015; Blondelle 2015; Yu 2015). These actions, which are mainly demonstration initiatives, aim at improving the on-field experience enhancing also the cooperation of industry and research institutions as well as to increase the awareness of end users. According to Carter and Wing (2013), about 8500 SOFC units for an overall power of more than 90 MW were installed in Asia, Europe and North America in the period 2009–2013; on the average the installed power almost doubled every year form 2010. From that analysis the Far East is leading the global market with both μ-CHP and high-power stationary applications. The achievement of economies of scale involving both manufacturing and capital costs may lead to the reduction of SOFC unit prices, which in turn could improve their market spread. It is worth noting that at the current stage, the deployment of SOFCs entails primarily economy, financial and industrial matters, regulatory and legislation issues as well as government subsides to leverage the potentialities of fuel cells for a cleaner environment. Although this text refrains from making economic analysis, it is easy to comment that FC technology is experiencing the same market trend of other high tech products that have been released to the market in the last century, and again the reduction of price has lead their large diffusion. When dealing with that issue two opposite questions may arise: Is the market leading to low price? Conversely, is the low price making a large market? The right answer is that a joint effort gathering together government, industrial and research institutions is required to push forward that technology toward a widespread diffusion. A study dealing with stationary fuel cells in distributed generation has been sponsored by the FCH-JU[1] and published in 2015 (FCDISTRICT 2010; Ene.field 2011; Ellamla et al. 2015); it reports on a coherent route toward the successful commercialization of stationary fuel cells in Europe with a comprehensive analysis of the great advancements seen in Japan and South Korea, as well as in the US. In the former countries a support to the large-scale diffusion of residential CHP systems is in place; on the other hand in the US the support is focusing more on industrial applications. The study shows the enhancements achieved through the commercialization of stationary systems, which has lead to a growth of the produced units resulting into a significant reduction of costs along with an improvement of the performance.

On the technical side fuel cells durability is the main challenge and only research and development efforts can lead to a stable and continuous improvement of the technology with an increase of fuel cell lifetime. It is known by the research community that the key role to improve and maintain stable FC performance is played by the materials used to build FC electrochemical components. Development of new materials and the

---

[1]The Fuel Cells and Hydrogen Joint Undertaking is a public–private partnership gathering at European level the European Commission, the fuel cell and hydrogen industry and the research institutions.

improvements of current ones are the traditional investigation areas with a great effort provided by the research community (Mahato et al. 2015). Considerable care is also given to other issues, such as cells and stacks design or the influence of mechanical strength on the material durability (Roehrens et al. 2015). Although material research can find the breakthrough solution to make a major step toward 2020 performance targets, from the industrial perspective an assessment of the technology is crucial to deploy commercial systems, which in turn may lead to the achievement of market's targets mentioned before. Therefore SOFC research is twofold: (i) find new materials, improve their use, enhance chemical processes to raise the performance; and (ii) guarantee the achievement of that performance on the systems, which must be installed on-field. The former objectives should be pursued through deep research on materials, whereas the latter may be achieved only if the technology is assessed and therefore stable. For successful deployment and long-lasting operations of SOFC, a major step is required toward the second objectives through methodological studies and the implementation of effective experimental and numerical techniques. Among others, some key functions such as control, diagnostic and prognostic as well as optimal stack design and periphery components sizing must be considered for their role in guaranteeing the attainment of performance targets. Models for design relying on reduced order mathematical formulations, but accurate and computationally fast, would support the optimization of stack, BoP and control strategies as well as the development of control and diagnosis algorithms. Either gray- or black-box models may represent the right framework to help development phases and system operations (Bove and Ubertini 2006; Wang et al. 2011). In this scenario, complex multidimensional models can also be exploited for the design of electrochemical components, as virtual fuel cell stack, providing data otherwise not available, or feeding the lower order model classes. A deeper integration of models and experiments as, e.g., in experimental design techniques, are envisaged for the enhancement of FC technologies. The scientific literature reports on few attempts on these areas though a quick turn toward a widespread use of new model approaches should be considered nowadays. The leverage of advanced industrial design methodologies—already exploited in other industrial areas—has to be promoted as well.

Although recent progresses have improved control strategies, conventional approaches may not be appropriate to maintain the control targets within feasible and safe operation domains for degraded or faulty system. By analogy with other mature industries, such as automotive, aeronautics and electric machines, an effort is required toward the objectives mentioned above. However a step ahead has to be made today introducing the future challenges, which are the integration of diagnostics, prognostics and advanced control to improve lifetime. Condition monitoring algorithms to evaluate the state of health (Marwala 2012) may give a great opportunity to improve both stack and components operations. The detection of malfunctioning parts as well as the quantification of the aging and their impact on system performance represent a step ahead toward the improvement of fuel cell lifetime. Moreover their integration with advanced control concepts (e.g., adaptiveness, robustness) can guarantee effective control to further optimize the performance and to mitigate the consequences of degradation and faults (Sorce et al.

2014; Polverino et al. 2015). The next generation of diagnostics, prognostics and advanced control devices and algorithms has to be developed by integrating, in a new framework, both models and dedicated experimental analysis. Also advanced measurements (e.g., electrochemical impedance spectroscopy) may be taken into consideration for on-board[2] implementation.

The above description deals with the exploitation of methodologies that should be applied to current SOFC systems where the stack, which represents the current status of the FC technology, is connected to the system peripheral components (e.g., blower, heat exchangers, postburner, prereformer, etc.) with a "standard" configuration. It is worth remarking that a new SOFC configuration, called integrated stack module (ISM), has been recently introduced by some manufacturers (Holtappels et al. 2005; Tsikonis et al. 2014). Such a system fits into a single adiabatic package all the hot components: the stack, the heat exchangers, the postburner, and the prereformer. It guarantees improved performance with respect to the standard configuration thanks to a better uniform temperature distribution among the hot components. However, the design, the control and the diagnosis of this ISM are more difficult with respect to the conventional system because of the heat exchange process occurring among the components. Again the challenge will be the model framework, which serves as a basis for the control and diagnosis algorithms development.

1.3 Benefits of Model-Based Approach for Control and Diagnostics

The key role of mathematical models in all engineering and industrial activities has increased in the last two decades. Notably, the design of any component either hardware or software has been enhanced through the use of models able to simulate its features. The central function of the models has become crucial with the increasing complexity of the systems to be designed, because of the limited capability of designers in accounting for all variables involved in that process. In this context the models may be seen as those algorithms able to reproduce the behavior (or the performance) of the system understudy. However, this statement does not give a complete picture of model functions and potentialities, which have become more enhanced and appealing for advanced use. Indeed models are nowadays embedded into the system itself to implement advanced functions, e.g., monitoring, control, diagnosis, maintenance scheduling, safety tasks, remote, or unsupervised operations.

Control and diagnostics of SOFC benefit from the use of models during design process, where the model reproduces the behavior of the SOFC stack and BoP (i.e.,

---

[2]Throughout the text on-board means any algorithm or device implemented on the system while working on real environment.

the plant). On the other hand, different models, which could be simpler than the plant model, may be either embedded into the control and diagnostic algorithms or run in parallel to the real plant. In the former case the models perform their distinctive features, e.g., to provide the right inputs to the plant actuators, to support an inferential tool for the fault detection process. In some diagnosis schemes the model running in parallel simulates the current system to monitor it for further status identification, e.g., normal, degraded or faulty. The above tasks that entail the use of one or more models, either running offline (e.g., design) or online (e.g., control, diagnosis, monitoring) belong to the broad category of model-based approaches (Bove and Ubertini 2006; Wang et al. 2011). For the sake of completeness, it is worth recalling that the control and diagnosis functions may be successfully implemented making use of non-model-based approaches, though in many practical applications mixed approaches would be the best choice. As an empirical rule, when a limited knowledge of the system to be controlled or diagnosed is available a non-model-based algorithm may be implemented. However, a selection of the most suited approaches has to be also made according to technical constraints, which may be imposed by both system and application boundaries, e.g., complexity of mathematical model, linear/nonlinear behavior, experimental data available, required accuracy, type and number of sensors installed, reliability, end-of-line manufacturing dispersion, maintenance, expertise of both engineers and operators, development time. From an industrial point of view this is a strategic trade-off to be solved according to company investments and business models, costs and product markets. The analysis of these issues is not herein focused, though few considerations will be made afterward to show how to select model-based or non-model-based approaches.

For control applications several techniques may be listed. The most widespread is the PID (proportional, integral, derivative) feedback controller, which is the simplest one. It does not require the embedding of any model, though an effective design needs a simulation model or an experimental campaign for the proper tuning of the algorithm. Other techniques, for either feedforward or feedback implementation, rely on maps to find the values of the controlled variables to be actuated; again such controllers need either models or experiments for their design. Artificial intelligence (AI) methods implementing computational tools such as neural networks, fuzzy logic, neuro fuzzy may also be applied for control purposes. These methods have received great attention in the last decade because of their capability in learning the process and providing the right control actions. Among others, the main advantages are their intrinsic nonlinear behavior and the multi-input multi-output (MIMO) features, which allow implementing complex nonlinear control strategies. AI methods feature input–output computational structures and—in principle—do not require the knowledge of the physical process to be controlled. On the other hand, large and exhaustive experimental data are required to instruct[3] the algorithms. Moreover, some AI algorithms have built-in features for adaptation

---

[3]In the artificial intelligence field the terms instruction and learning stem from the analogy of the computational process with that of the human brain behavior.

and therefore might be worth using to account for the drift of performance due to aging or other causes, which are not easily predictable (e.g., degradation). Neural network controllers share some common problems with neural network models, which will be recalled in Chap. 2 when dealing with black-box models for control applications (Marra et al. 2013; Sorrentino et al. 2014). Among others, the extensive experimental data required may limit the implementation and the upgrade of these algorithms, thus reducing their flexibility; another issue is the lack of stability that might affect such controllers because of their high nonlinearity. For fuzzy logic algorithms a distinctive feature is the possibility to set up the computational structures, either model or controller, by using linguistic relationships to code heuristic knowledge via simple reasoning processes. Therefore, the fuzzy logic algorithms are built through a rather long sequence of multiple elementary semantic rules that implement logic operations, which are able to describe the complex behavior of the system to be controlled. The main limitations are (i) their inherent simplicity, which could not support the description of complex physics, and (ii) the inaccuracy due to an incomplete knowledge of the process being reproduced. For a comprehensive review of the most effective methodologies for fuzzy logic algorithms the book of Babuška (1998) is a good reference for practiceners. Neuro-fuzzy algorithms exploit the training techniques of neural networks to tune the fuzzy functions, thus improving the performance of basic algorithms by merging mathematical techniques and reasoning-based approaches (Sandhu and Rattan 1997).

For diagnostic applications a wide set of non-model-based approaches is available; among others, artificial intelligence, statistical (e.g., Bayesian networks, principal component or Fischer discriminants) and signal-based (e.g., Fourier analysis, wavelets) methods can be applied to fuel cells. It is worth to note that non-model-based approaches discriminate between nominal (i.e., safe) and faulty operations by using the system knowledge stored into each algorithm. That knowledge is built making use of large experimental database populated by measures gathered during both nominal and faulty operations. Therefore the development of such algorithms requires comprehensive experimental campaigns to be performed during nominal and off-design or faulty operations. For an exhaustive review on non-model-based algorithms the paper of Zeng et al. (2013) gives a thorough description of those techniques for PEM fuel cells; their application to SOFC or other fuel cell technologies is straightforward.

As far as model-based diagnostics concerns, it is worth recalling the basic principles of such an approach. The main objective of a diagnostic algorithm is to detect and isolate (FDI) undesired states (i.e., faults) within the system under study (e.g., both stack and BoP components of the SOFC system). The algorithm exploits a process model to monitor the system during its operations in real time. A set of data is also measured and treated together with monitored variables to obtain insightful indicators (i.e., features) of system's state (Isermann 2005). The features are then compared against reference thresholds whose values are suitably tuned to take into account uncertainties (e.g., model inaccuracy, measurement noise) and the necessity to detect incipient faults. The comparison of the features to the thresholds allows the generation of analytical symptoms, which indicate whether an undesired event is occurring or not. The arising of a symptom states that the behavior of the related variable attains abnormal values, thus confirming the occurrence of a fault whose type has to be determined. Then to isolate it, a reference set of information has to be used to correctly locate the fault on the system. The isolation may be performed making use of several approaches, among others, a Fault Signature Matrix (FSM) (Arsie et al. 2010) can be developed on the bases of heuristic or analytical Fault Tree Analysis (FTA) as reported by Polverino et al. (2015). It is worth to remark that a successful model-based diagnosis requires a monitoring model to simulate either a single process or the entire system in real time with the required accuracy. As reported above to successfully isolate a fault a direct measure of relevant variables must be taken. Therefore BoP faults can be easily identified through the direct measure of conventional variables, such as pressure temperature, mass flow rate, current drawn by the actuators (e.g., blower, valves); in some cases concentrations may be also measured via advanced sensors (e.g., automotive-derived O_2 sensor), as adopted by Fadeyev et al. (2015). On the other hand to discriminate stack faults, models of in-stack phenomena should be implemented to simulate those processes that may be affected by faults; thus electrochemical, electrical, mechanical, thermal, and fluid dynamic models should be derived to monitor the stack. Although these phenomena are difficult to simulate with fast models some solutions can be found by implementing, e.g., reduced order models. The most critical part of the FDI process is to gather direct measures of in-stack variables due to the difficulties to install sensors within the stack and the lack of sensors providing direct measures of, e.g., electrochemistry-related variables. Therefore indirect information must be derived from the conventional set of sensors installed on the BoP as described in Chap. 4. However other techniques may be envisaged to indirectly measure those variables, among others, the Electrochemical Impedance Spectroscopy if implemented on-board may give a valuable contribution to improve stack diagnostics. Such an approach has been implemented for PEM systems (D-CODE 2015) and could be applied for SOFC as well.

A further issue dealing with stack performance degradation is becoming more and more relevant today. Indeed, the knowledge of the actual status as well as the prediction of its future evolution (i.e., lifetime forecast) are seen as advanced diagnostic functions to be embedded within any fuel cell system. These features may support all actions to be taken for stack/systems damage prevention as well as maintenance plans (predictive maintenance) with a direct impact on its availability, duration and therefore costs. The development of prognostic tools, able to predict degradation course and stack lifetime, entails the development of new models or the use of experimental data gathered during massive campaigns performed for either degraded or aged conditions. The research on SOFC prognostics has proposed few solutions so far and new approaches must be implemented at both methodological and phenomenological levels. This can be considered the next frontier of the SOFC research to be achieved in the coming years. Methodological approaches can be easily borrowed from other areas where prognostics and health management techniques have already been implemented (e.g., automotive, aeronautics, nuclear). On the other hand, the main issue for SOFC will be the development of fast models to simulate the impact of leading degradation phenomena affecting SOFC performance. Among others, the primary phenomena concerned are sulpur poisoning; carbon deposition; Nickel agglomeration, coarsening and reoxidation; high fuel utilization; electrolyte cracks; porosity; Chromium and Si poisoning; contact degradation and delamination.

1.4 Literature Survey

In the following is given a brief review of the main works dealing with models for control and diagnostics. The literature analysis reports a large list of control-oriented models whereas few works concern with diagnostic-oriented ones. All models aim at control and diagnostic applications with two main objectives: the first concerns with the design of both control and diagnostic strategies, whereas the second deals with the on-line monitoring of SOFC systems. In all cases a reduced computational time is a must for, e.g., optimization purposes or analysis of scenarios; on the other hand the models must be precise enough for an accurate description of the system dynamics. Usually one-dimensional (1D) or zero dimensional (0D) models, i.e., lumped, combine together these features. The on-line monitoring is a key part of both diagnosis and control algorithms, as described in the following chapters; it requires models running faster than real time, in such a way that they can simulate the system process in parallel to the real one: data-driven (or black-box) models are the most suitable for these applications. In some cases, thanks to the availability of high computational power, 0D dynamic models may be implemented on board for real-time monitoring as well. An interesting feature is the use of 1D and 0D models both for the design of control and diagnosis strategies and for the generation of virtual data exploited to develop black-box models. This latter application deals with the hierarchical approach that proved useful for several engineering implementation and is extensively explained in Chap. 2. Although today conventional computers perform high computational speed, multidimensional models (i.e., 1, 2, 3D) are not yet suitable for the purpose of either design or on-line applications, such as control and diagnostics; therefore, this broad area has not been herein reviewed. However, the foreseen development of both computing technology and numerical methods will surely let multidimensional models entering into the arena of control and diagnostics in a short while. The reader interested into 2D and 3D models is addressed to the paper of Wang et al. (2011) who reports a thorough review on solid oxide models that was assembled from the viewpoint of researchers dealing with diagnostic applications. For the purpose of introducing the main model structures and the most relevant works available in the literature, the following part of the paragraph describes 1D models, which are appropriate for design of both control strategies and diagnostic algorithms, then 0D and black-box models are reviewed for their useful implementation within model-based control and diagnostic algorithms. To give a glance at an active research area that is receiving great attention for its relevance in the frame of both advanced control and fault detection, a short survey of models approaches suitable for degradation analysis is also reported. Finally a quick summary is given on a more complex approach dealing with equivalent circuit model (ECM) whose implementation for control and diagnostics requires the measurement of electrochemical impedance spectroscopy (EIS).

1.4.1 1D Models

The design of control and diagnostic algorithms may be performed making use of both stationary and dynamic 1D models, which aim at analyzing temperature and gas concentrations distribution within either single cell or stack. Temperature levels and gradients inside the stack are the main issues in the control of SOFC systems because of the thermal stresses induced on solid material, which could cause, e.g., delamination, mechanical failures, as well as electrochemical degradation.

Bao and Bessler (2012) proposed an interesting 1D + 1D (quasi-2D) model whose main feature is the decoupling of the physical phenomena in the two main directions. The main equations (mass and energy transport) are integrated along the flow direction whereas only energy and heat transfer are accounted in the transversal direction. The model was built by coupling both semiempirical and analytical submodels to simulate SOFC performance in steady-state conditions. In this work, two different cases were analyzed, i.e., isothermal and non-isothermal. The model is fairly fast to complete in seconds the simulation with 20 or 50 along-the-channel control volumes. Cheddie and Munroe (2007) developed a dynamic 1D model for real-time simulation that was derived from previous 0D real time one. The results presented show that a model with 21 nodes exhibits the best balance between computational time and accuracy. Such a model required few milliseconds of computational time for a single timestep. The computational time is comparable with that of a 0D model, but the cell is considered adiabatic and no heat transfer with the external environment is modeled. Sorrentino et al. (2008) proposed a hierarchical modeling approach to derive a control-oriented lumped model of planar solid oxide fuel cells, particularly developed by exploiting virtual experiments generated via a steady-state 1D model of single cell operation. The proposed model is capable of accurately simulating temperature and voltage dynamics as function of the main operating variables (i.e., current density, fuel and air utilizations, inlet and outlet temperatures). The 1D model was extended to simulate the performance of a planar radial SOFC by Marra et al. (2015a). In the work a hierarchical approach was adopted first for the identification and validation of the 1D model and then to identify a control-oriented model based on Neural Networks (NNs). The 1D model was identified and validated by using the results of a 3D model available in literature (Andreassi et al. 2007). The control-oriented model simulates the stack voltage and is suitable to be implemented into control and diagnostic structures for real-time monitoring. Kang et al. (2009) proposed a simplified 1D dynamic model of a planar co-flow adiabatic Direct Internal Reforming

(DIR) SOFC. The results were satisfactorily compared to those obtained from a detailed 1D SOFC model and the computational time was limited only to few seconds. To develop a simpler 1D dynamic model, Qi et al. (2008) proposed an approximate analytical solution for the space distribution of reacting gas flow. By this method, the 1D dynamic SOFC model was reduced by assuming the space coordinate as a parameter, thus the model structure became a nonlinear state space one, made of a set of ODEs and is therefore suitable for control design. Wahl et al. (2015) presented a dynamic quasi-2D model of an SOFC system with anode off-gas recycling, which consists of stack and balance of plant including thermal coupling among BoP components. Zhang et al. (2015) proposed a dynamic hybrid model of a 5 kW SOFC by merging 1D and 0D models. The 1D model was adopted to simulate stack temperature field distribution in the gas flow direction as well as heat exchangers to account for an accurate representation of heat transfer processes. Other SOFC components, including a burner, were modeled with a lumped approach (i.e., 0D) to speed up the simulation. As far as tubular SOFC concerns, Barzi et al. (2009) simulated the start-up behavior making use of a 2-D dynamic model, based on finite volume method to solve the nonlinear equations. Such a model showed high computational burden due to the coupled description of the physics with high accuracy in predicting temperature distribution in the bulk material along the two directions. Thanks to its dynamic features the authors claimed the application of the model for the design of control strategies suitable for transient maneuvers management.

1.4.2 0D Models

The greater part of the literature on control and diagnosis deals with 0D dynamic models for stacks and systems. Such models, as reported before, are exploited both for on-line monitoring and for the design of control and diagnostic strategies. These models are suitable when the main characteristics and the performance of the system are known, e.g., when a prototype is already available. Thus, the lack of some physical information may be overcome by introducing data extracted from experiments or from complex multidimensional models, whose outputs may be used as virtual experiments in the frame of hierarchical modeling. Therefore, 0D models merge a simplified phenomenological description together with practical or empirical information. For the sake of completeness, it is worth to report that the main equations of such models are still based on first principles (e.g., conservation of mass and energy). When spatial variations are not accounted, a 0D model is called lumped and, in some cases, the term phenomenological is considered to remark that the main phenomena are represented with an integral approach. Thus the mathematical formulation results to be easy to handle with respect to an equivalent full-scale physical model. In the case of lumped ones, the single elements of a system (e.g., stack, compressors, heat exchangers, fuel reformer, partial oxidizers, contaminant removal apparatus, etc.) are simulated through different submodels to further simplify the mathematics (Bove and Ubertini 2006; Sorrentino and Pianese 2009a, b).

The development of a 0D model starts from the a priori knowledge of the process to derive a set of mathematical equations—usually ODEs—, which describe the time evolution of the system. Afterward, the calibration of the model is performed through the identification of unknown parameters; then, to better appraise its accuracy (Sohlberg 2003), the model outputs must be compared against experimental data acquired during independent experimental campaigns. The entire process allows meeting the trade-off between satisfactory accuracy and affordable computational time as reported by Bhattacharyya and Rengaswamy (2009) in their detailed review of SOFC dynamic models. In the following, some 0D models are selected from the literature with attention to their features and solutions proposed to make them consistent with the requirement of fair 0D models for either control or diagnostics. It is worth to note that most of the literature covers control-oriented models, whereas little works deal with diagnostics for on-board applications, having this latter area received attention from the SOFC community only after 2005.

Costamagna et al. (2001) described a hybrid turbogas hybrid system where the SOFC/GT was simulated by a 0D model approach. The governing equations were written as macroscopic integral balances between inlet and outlet fluxes of mass and energy in each system component. Such a pioneering work can be considered as one of the first references for control oriented models. Zhu and Tomsovic (2002) analyzed the load following properties of a SOFC/GT system. They adopted the SOFC model of Padullés et al. (2000) and developed dynamic models for the rotating parts and GT system components, which should be considered as part of the BOP for an SOFC/GT assembly. Ferrari et al. (2004) and Magistri et al. (2006) exploited a dynamic model to perform transient analysis of a hybrid turbogas system based on SOFC. This configuration mainly consisted of three parts: the stack, the anodic recirculation system with fuel feeding and the cathodic side (air side) with a gas turbine plant and heat exchangers. An interesting lumped approach was followed by Sedghisigarchi and Feliachi (2004a, b) for control and stability enhancement of SOFC-based distributed generators; in such an approach, the average cell temperature was assumed as a state variable. Xi et al. (2007) used a minimum Gibbs free energy approach to determine the flow gas composition to simplify the calculation of the mass balance within the SOFC. This allows developing a control-oriented model that achieves valuable trade-off between model accuracy and computational burden. Mueller et al. (2007) developed a dynamic model of an SOFC/GT system. The model was exploited to analyze the system at several operating conditions and to design the control strategy. The analyses indicate that, for SOFC/GT hybrid plants it is more performant to control system power by manipulating fuel cell current, changing the anode fuel flowrate, and to control fuel cell voltage. Gaynor et al. (2008) studied the fuel starvation problem in SOFC and suggested three methods to prevent it through control actions: use of a rate limiter on the load, changes on the fuel flow controller and reference governors. These latter use a dynamic lumped model to predict the future response of the system to set the reference control commands, instead of using fixed set points.

A lumped model was also applied for the transient modeling of a tubular SOFC by Hajimolana and Soroush (2009); they developed suitable strategies aimed at controlling voltage and cell-tube temperature by properly acting on both temperature and pressure of the inlet air flow. Sorrentino and Pianese (2009b) presented a dynamic model of an SOFC unit for CHP applications with a planar co-flow stack surrounded by air blower, regulating pressure valves, heat exchangers, prereformer and postburner. As a consequence of low thermal dynamics characterizing SOFCs, a lumped capacity model was successfully proposed to describe the coupled response of both stack and heat exchangers to the load change. The same model was also applied for the control of an SOFC during cold-start and warmed-up transient operations (Sorrentino and Pianese 2011). Nanaeda et al. (2010) proposed a dynamic model of an SOFC-based CHP system to understand system operating limits and improve flexibility taking into account the constraints imposed by end-user needs and system efficiency. The model is based upon the physics, chemistry and electrochemistry that govern the system, considering conservation of mass and energy. The submodel of each system component was developed individually and integrated into a global model, which was exploited to analyze the dynamic behavior of the entire CHP system and to develop the control strategies.

A 0D model for the simulation of SOFCs based micro-cogenerative power system fed by natural gas is described by Arpino et al. (2013), who focused their work on the control logic implemented on-board. The model was validated and then used to investigate the generator performance under different operating conditions; it is worth noting that such a model was also applied for Design of Experiment (DoE) activity. Barelli et al. (2013a) proposed a dynamic model of an SOFC-based CHP system to evaluate the thermochemical operating conditions. Their goal was the optimal sizing of the main plant components, to guarantee suitable system inertia and to evaluate its global performance. From the same group (Barelli et al. 2013b) a work on SOFC/GT was done for a dynamic analysis during part load operation. Again, a lumped dynamic model of the system was used assuming a constant operating temperature.

Also SOFC-based Auxiliary power units (APU) have received great attention in the last decade for their potential use on, e.g., trucks, thanks to their fuel flexibility and potential low emissions features. Although referenced as APU, these systems use the exhausted heat for cabin heating and cooling if combined with trigeneration devices. Among others, Lu et al. (2006) modeled an SOFC APU for trucks, made of one SOFC stack, two heat exchangers, one combustor, one controller and power electronics. Murshed et al. (2007) modeled a planar SOFC system with BOP components including reformer, burner, and heat exchanger. Sorrentino and Pianese (2009a) developed a dynamic model for an APU system accounting for the coupled dynamics of all components that allowed for a thorough control design of warming up phases (Sorrentino and Pianese 2011).

In the frame of the FCH-JU-funded EU project DIAMOD (2014), Vrečko et al. (2015) used the lumped dynamic model of Sorrentino et al. (2008) to design a feedforward–feedback controller for the stack temperature of an SOFC Stack. The algorithm shows good performance in controlling the temperature and its easy tuning makes it feasible for on-board implementation. Within the same project, Marra et al. (2015a, b) developed a lumped model to simulate the temperature and anode gas composition at the outlet of an SOFC stack with anode off-gas recirculation. The model accurately reproduces real data for transient maneuvers of a 10 kW SOFC system; it may serve as state estimator for model-based algorithms to be embedded within either controllers or diagnostic tools. Starting from the fault tree analysis of an SOFC system (Arsie et al. 2010), Polverino et al. (2015) used the lumped dynamic model of Sorrentino and Pianese (2011) to develop a procedure for the design of a diagnostic algorithm. A fault-to-symptoms dependency analysis was carried out making use of a detailed simulation of 5 faults occurring in the system (one in the stack and four at BOP level). A thorough study was performed accounting for the influence of the operating conditions, the faults magnitude and the level of the thresholds. Sorce et al. (2014) presented a lumped dynamic model of a laboratory-size SOFC system. After validation the model was used to simulate four classes of system faults, i.e., air leakage, fuel leakage, SOFC degradation and reformer degradation. The results of the fault analysis represent a basis for the development of a fault detection and isolation (FDI) tool based on pattern recognition techniques. Greco et al. (2014) employed the model developed by Sorce et al. (2014) to analyze reformer faults by comparing experimental and modeling results. Fault maps were proposed as basis for the development of FDI tools.

1.4.3   Black-Box (Data-Driven) Models

Contrary to physical models (either 1D or 0D), black-box ones are not based on physical equations but databases with experimental data and therefore are also called data-driven models. The lack of physical relationships within the model structure forces the use of a large set of input–output pairs to be used for the identification of model parameters. Therefore for accurate modeling, the experimental data must contain meaningful information on the physical behavior of the system. These models range from classical regression-based approaches to artificial intelligence-based ones (e.g., Neural Network), which have been demonstrated to be suitable for nonlinear systems (Patan 2008). However, such models require a very large amount of data (i.e., training examples), which should describe the behavior of the system; therefore, the experimental burden may become excessive. Although the experimental load is the main drawback of artificial intelligence-based modeling techniques, their intrinsic high accuracy represents the most attractive characteristic. These two opposite features lead to the main trade-off to deal with when approaching the modeling problem to be solved. On the other hand, the accuracy of the model is guaranteed only if the simulated system is strictly the same as that used to generate the input–output data for the parameters identification.

Milewski and Swirski (2009) implemented the same NN structure of Arriagada et al. (2002) to simulate the SOFC behavior. This SOFC model predicts the output cell voltage making use of 9 inputs: current density; cathode inlet O_2 and N_2 flow densities; anode H_2 and He flow densities; anode thickness; anode porosity; electrolyte thickness and electrolyte temperature. The accuracy achieved is very high and it is extremely useful for rapid calculation of SOFC performance under dynamic operations. The limit is represented by the low generalizability of the model to simulate SOFC behavior with different geometries or material. In order to ensure SOFC system safe operation and lifetime, Wu and Zhu (2011) analyzed the control logic problem of a hybrid SOFC/GT system and implemented a mathematical model with multi-loop control strategy making use of a dynamic radial basis function (RBF) neural network and an adaptive PID controller. The analysis of the temperature dynamics in an SOFC/GT system, to support temperature control strategy design and analysis, was performed by Wu et al. (2011), who developed a dynamic model of the system, including the thermal coupling among different components (i.e., fuel cell stack, combustion zone and balance of plant components). In order to identify model parameters the authors adopted a least squares support vector machine (LS-SVM) algorithm, based on an improved particle swarm optimization (PSO) approach. This methdology was compared with the back propagation neural network approach showing higher prediction accuracy and faster convergent speed. Within the project DIAMOND (2014), Dolenc et al. (2015) used a data-driven approach to design a set of soft sensors for on-line estimation of oxygen-to-carbon ratio and for minimal and maximal stack temperatures. These sensors can be implemented on-board to feed the controller with simulated data instead of using physical sensors.

For diagnosis-oriented SOFC stack simulations a deep study was performed within the FCH-JU-funded EU project GENIUS (2009). Several black-box approaches were exploited ranging from regression-type (Guida et al. 2015) to neural networks. In the same project Wang et al. (2012) proposed a Bayesian Network, whereas Marra et al. (2013) developed multilayer perceptron Neural Networks. This latter model includes the time as input variable and is able to simulate the SOFC stack voltage in presence of degradation; it is therefore suitable for advanced on-line monitoring, control and diagnostic algorithms. Dynamic neural networks were also tested satisfactorily for the transient simulation of SOFC with single and multi-stacks configurations (Sorrentino et al. 2014).

1.4.4   Models for Diagnosis and Degradation Monitoring

Although on-board diagnostics is considered as one of the key functions able to support FCS optimal performance achievement and to enhance reliability, a systematic attention to its development has been given only after 2005. Most of the open literature reports on studies focusing on diagnostic analysis oriented toward the identification of stack faults and failure during laboratory operations. The major attention has been given to study the behavior of FC with respect to new materials with the objective of improving primarily cell performance. The concept of on-board diagnostics for fault detection and isolation as well as for health monitoring has captured more and more attention by both research and industry communities. It is now considered as an important asset to foster the enhancement of SOFC systems as marketable product. Therefore, model-based diagnostics has become an area of continuous investigation supported by public and private funds and the literature has grown reporting on several methodologies already exploited by other mature industrial sectors and made available for all type of fuel cells. Remarkable attention is given today to nondimensional models. Indeed black-box or lumped models are well suited for those applications thanks to their monitoring capabilities, which mainly concern with computational speed, accuracy and implementability on real systems. The fast modeling of stack degradation phenomena is receiving more and more attention in view of further enhancement for on-board monitoring and diagnostic functions. Indeed they could support (i) the forecast of stack remaining useful life (RUL); (ii) the advanced control whose parameters may be adapted to optimize performance or minimize the degradation; and (iii) the detection of stack faults. This latter can be implemented only if relevant measures or indirect information on in-stack phenomena are available.[4] Toward this objectives advanced measurement systems are required, among others, electrochemical impedance spectroscopy (EIS) is a valid candidate once its on-board implementation will be completed, as pursued by the project D-CODE (2015) for PEMFC. Another technique based on total harmonic distortion analysis could be worth using as shown for PEMFC by Ramschak (2009).

In the closure of this chapter an overview of monitoring models for state of health and degradation analysis together with a short recall of EIS-based approach are given. Stack performance are usually affected by different mechanisms involving single o mutual component effects, such as electrode delamination, thermochemical and thermomechanical stresses and electrode poisoning, just to mention a few. However, distinct causes may induce equivalent effects on the stack performance. In such a case, it is quite hard, if not unattainable, to univocally identify and isolate a specific mechanism through the analysis of conventional thermodynamic, fluid dynamic and electrical measured data (e.g., pressure, temperature, output voltage) only. Therefore more advanced diagnostic techniques are needed to enhance the capability of diagnostic tools; among others, it is worth mentioning electrochemical impedance spectroscopy.

To describe stack degradation phenomena Larrain et al. (2006) developed an SOFC repeated elements model in order to analyze stack degradation due different phenomena. The considered degradation mechanisms are mainly related to the electrodes (e.g., presence of impurities, microstructural changes, conductivity decrease, etc.) and the interconnects (e.g., thermal stresses, oxidation, etc.). Moreover, anode reoxidation was also analyzed with respect to operating conditions and design choices. The same objective was achieved by Virkar (2007), who developed an electrochemical SOFC stack model able to simulate the degradation

---

[4]As detailed in Chap. 4 any effective fault detection algorithm requires a direct measure of the phenomenon relevant to the fault to be diagnosed.

induced by the increase in the resistance of an isolated cell (or few cells). The causes of such a mechanism could be caused by several phenomena, such as the formation of local hot spots—which can induce the change of material properties and microstructures—fuel or oxidant nonuniform distribution, seals degradation or electrode delamination induced by thermal cycling. Barelli et al. (2013c) proposed an exhaustive analysis of several degradation mechanisms affecting SOFCs and the diagnostic procedures currently available in literature for their detection. For instance, the authors accounted for electrode delamination, thermal stress, carbon deposition and sulfur poisoning. To detect the considered degradation phenomena, they showed how effective is the use of EIS to identify a delaminated cell and the attainment of a probability analysis to define the probability of failure at different thermal conditions.

Although a significant effort is currently performed to investigate and detect fuel cell degradation mechanisms, a deeper study on the influence of all the SOFC system components (i.e., stack and BOP) on the system behavior during normal and faulty condition is still lacking. Furthermore, only a limited number of authors developed mathematical models to simulate systems malfunctions, faults or failures. Notably, the paper of Escobet et al. (2009) deals with PEM FC systems and is one of the first who reported on the application of well structured model-based diagnostics to identify six faults occurring in the BoP. The work of Riascos et al. (2008) is worth to be mentioned for its valuable scientific contribution dealing with a dedicated experimental activity performed on a PEM system during both nominal and faulty operation. On this line, Ingimundarson et al. (2008) presented a model-based approach to detect hydrogen leaks in PEM fuel cell stacks. They developed a lumped model based on filling–emptying approach to compute the hydrogen leaked mass flow rate. In order to fulfill model validation, these quantities were simulated through the proposed model and compared to those obtained during dedicated experiments. To achieve detection, they also set an adaptive threshold, function of natural leak, pressure and temperature. For fault detection occurring in SOFC systems the works of Arsie et al. (2010), Greco et al. (2014), Sorce et al. (2014) and Polverino et al. (2015, 2016) were already described above.

Other diagnostic approaches exploit electrochemical impedance spectroscopy to deepen the analysis on system state of health and degradation behavior. Electrochemical impedance spectroscopy is used to derive the impedance of an electrochemical device at different frequencies by superimposing a small voltage/current AC sinusoidal perturbation onto the voltage/current operating point. Then a set of impedances is derived as function of the frequency (i.e., impedance spectrum). The basic idea behind EIS is to identify the response of the device after a continuous set of sinusoidal perturbations imposed on the system at different frequencies (from kHz to tenths of Hz). Therefore the information brought by the EIS is very rich thanks to the dynamic nature of the signal. For the purpose of monitoring and diagnosis, the time varying of the spectrum can provide some indications about the evolution of the electrochemical device. By combining EIS and fuel cell equivalent circuit model quantitative data for diagnostic purpose may be derived. Equivalent circuit models are built with electrical elements such as a parallel RC circuit, in the simplest case (Singhal and Kendall 2003). Each discrete element of the model is supposed to have corresponding physical meanings and thereby a straightforward interpretation of the FC status can be made. In case of complex model structures, model parameters are evaluated through parameter identification techniques (Petrone et al. 2013; Wang et al. 2011). In some cases, the parameters can intuitively provide the information about the processes occurring inside the system; in other cases, however, a complex inference should be carried out. This phase has a function equal to that of mapping, i.e., establishing a correlation between the identified model parameters and the investigated characteristics of the system (Huang et al. 2007). Once the values of these elements (parameters) are identified in various operating condition, they can be used as an estimated data set for fitting physical models. This modeling method is thoroughly described by Leonide et al. 2009. Among others the work of Gazzarri and Kesler (2007) proved EIS being a useful method to diagnose several degradation phenomena. Also Huang et al. (2009) developed a diagnostic algorithm based on a metal supported SOFC equivalent circuit model; they investigated on the degradation phenomena by means of a dedicated experimental activity.

1.5 Chapter Closure

Fuel cells will contribute substantially to the greening of energy conversion technologies for both stationary and transportation applications as well as for those relevant to small and niche markets (e.g., APU, portable equipment). Many installations in North America, Europe and in the Far East have demonstrated the feasibility of SOFC, primarily for CHP uses; this has been recognized by all public and private stakeholders, which support research, industrialization and deployment actions. Moreover, the public opinion is today aware of the potentiality of fuel cell technology thanks to the appearance on the market of fuel cell powered cars. This has indeed confirmed that fuel cells are marketable "goods" rather than promising devices converting hydrogen into electric energy. Nevertheless reliability must be further improved to guarantee secure and long time operations. Therefore, research and technology advancements are required to enhance all components from its core (i.e., materials) toward ancillary devices (e.g., blower, power electronics) and their clever integration and management. In this process, control and diagnostics play a leading role, as well as prognostics that is however at its early stage.

It is worth recalling that models and both control and diagnostic algorithms presented in the book must be implemented within an electronic control unit and their performance are "responsible" for the system behaviour. Therefore accuracy and computational time are the foremost features to be guaranteed. Throughout the book these two issues are the main references when dealing with model selection, control strategy design and diagnostic tool building. Moreover, models and algorithms for control and diagnostics are developed for SOFC systems that are already built, thus experimental data could be available and exploitable to improve models and computational tools performance.

This book aims at providing the basics of modeling for the design of control strategies and diagnostics algorithms and is split into three main chapters. Chapter 2 introduces the modeling framework for SOFC systems based on hierarchical relationships among model approaches whose purpose is to solve the trade-off between computational speed and accuracy. Several modeling techniques are presented for both steady-state and dynamic simulations along with the main practical concepts for their proper selection and successful implementation. The chapter emphasizes the interrelations among available experimental data, accuracy and computational time. Chapter 3 describes the application of the hierarchical modeling structure for control strategies development; two examples of SOFC-based systems are given, namely truck APU and CHP. A multilevel control architecture is described for the purpose of achieving the targeted performance within an integrated structure combining both feedforward and feedback strategies. Notably, the operating load is guaranteed together with the proper air and fuel inlet flows to keep the thermal gradients across fuel cell length within the constraints imposed by material properties. The dynamic feature of the stack model facilitates the development of control strategies for cold-start and shut-down maneuvres limiting the temperature derivative. The last chapter deals with model-based diagnostic tool development whose objective is the detection and isolation of faults in stack and balance of plant components. This chapter gives an overview of the main concepts behind monitoring, fault detection and isolation functions. A focus is also given to the setting of threshold levels and its link with the trade-off between fault missing and false alarm, whose solution plays a central role when scheduling maintenance plans. The fault identification problem is addressed through the design of the fault signature matrix, which links faults and symptoms. In this chapter the conventional heuristic approach for the fault three analysis is described to support the signature matrix building along with a more analytical approach that exploits a model to simulate the behaviour of the system under normal and faulty operation. This latter approach is able to link in a quantitative manner the correlation among faults and monitored variables. Two case studies are proposed for system faults occurring in the air blower and in the prereformer heat exchange surface.

References

Aguiar P, Adjiman CS, Brandon NP (2004) Anode-supported intermediate temperature direct internal reforming solid oxide fuel cell. I: model-based steady-state performance. J Power Sources 138:120–136

Aguiar P, Brett DJL, Brandon NP (2007) Feasibility study and techno-economic analysis of an SOFC/battery hybrid system for vehicle applications. J Power Sources 171:186–197

Andreassi L, Rubeo G, Ubertini S, Lunghi P, Bove R (2007) Experimental and numerical analysis of a radial flow solid oxide fuel cell. Int J Hydrogen Energy 32:4559–4574

Arpino F, Dell'Isola M, Maugeri D, Massarotti N, Mauro A (2013) A new model for the analysis of operating conditions of microcogenerative SOFC units. Int J Hydrogen Energy 38:336–344

Arriagada J, Olausson P, Selimovic A (2002) Artificial neural network simulator for SOFC performance prediction. J Power Sources 112:54–60

Arsie I, Di Filippi A, Marra D, Pianese C, Sorrentino M (2010) Fault tree analysis aimed to design and implement on-field fault detection and isolation scheme for SOFC systems. In: Proceedings of the ASME 2010 eighth international fuel cell science, engineering and technology conference, FuelCell2010, June 14–16, 2010, Brooklyn, New York, USA, FuelCell2010-33344

Babuška R (1998) Fuzzy modeling for control. Kluwer Academic Publishers, Boston

Bao C, Bessler WG (2012) A computationally efficient steady-state electrode-level and 1D + 1D cell-level fuel cell model. J Power Sources 210:67–80

Barelli L, Bidini G, Gallorini F, Ottaviano A (2011) An energeticeexergetic comparison between PEMFC and SOFC-based micro-CHP systems. Int J Hydrogen Energy 36:3206–3214

Barelli L, Bidini G, Gallorini F, Ottaviano A (2013a) Design optimization of a SOFC-based CHP system through dynamic analysis. Int J Hydrogen Energy 38:354–369

Barelli L, Bidini G, Ottaviano A (2013b) Part load operation of a SOFC/GT hybrid system: dynamic analysis. Appl Energy 110:173–189

Barelli L, Barluzzi E, Bidini G (2013c) Diagnosis methodology and technique for solide oxide fuel cells: A review. Int J Hydrogen Energy 38:5060–5074

Barzi Y, Ghassemi M, Hamedi M (2009) Numerical analysis of start-up operation of a tubular solid oxide fuel cell. Int J Hydrogen Energy 34(4):2015–2025

Bhattacharyya D, Rengaswamy R (2009) A review of solid oxide fuel cell SOFC dynamic models. Ind Eng Chem Res 48:6068–6086

Blondelle J (2015) Fuel cells & hydrogen—EU level update.In: IPHE—23rd Steering Committee meeting, Wuhan, China, 27–28 May 2015. http://www.iphe.net/docs/Meetings/SC23/European %20Commission_SC23.pdf. Cited 23 Dec 2015

Blum L, Meulenberg WA, Nabielek H, Steinberger-Wilckens R (2005) Worldwide SOFC technology overview and benchmark. Int J Appl Ceram Technol 2:482–492

Blum L, de Haart LGJ, Malzbender J, Menzler NH, Remmel J, Steinberger-Wilckens R (2013) Recent results in Jülich solid oxide fuel cell technology development. J Power Sources 241:477–485

Bogicevic A, Wolverton C, Crosbie GM, Stechel EB (2001) Defect ordering in aliovalently doped cubic zirconia from first principles. Phys Rev B 64:141061–1410614

Bove R, Ubertini S (2006) Modeling solid oxide fuel cell operation: approaches, techniques and results. J Power Sources 159:543–559

Campanari S (2001) Thermodynamic model and parametric analysis of a tubular SOFC module. J Power Sources 92:26–34

Carter D, Wing J (2013) The fuel cell industry review 2013. Fuel Cell Today, Royston, UK

Cheddie DF, Munroe NDH (2007) A dynamic 1D model of a solid oxide fuel cell for real time simulation. J Power Sources 171:634–643

Cho HJ, Park YM, Choi GM (2011) Enhanced power density of metal-supported solid oxide fuel cell with a two-step firing process. Solid State Ionics 192:519–522

Christiansen N, Kristensen S, Holm-Larsen H, Larsen PH, Mogensen MB, Hendriksen PV, Linderoth S (2005). Status and recent progress in SOFC development at Haldor Topsøe A/S and Risø. In: Singhal SC, Mizusaki J (eds) Proceedings of cells, stacks, and systems, vol 1. Electrochemical Society, Incorporated, Pennington, pp 168–176

Costamagna P, Magistri L, Massardo AF (2001) Design and part-load perfor-mance of a hybrid system based on a solid oxide fuel cell reactor and a micro gas turbine. J Power Sources 96:352–368

D-CODE (2015) DC/DC Converter-based Diagnostics for PEM systems, project funded by the European Community's Seventh Framework Programme (FP7/2007–2013) for the Fuel Cells and Hydrogen Joint Technology Initiative under grant agreement n° 256673. http://www.d-code.unisa.it. Accessed 18 Aug 2015

DESTA (2010) Demonstration of 1st European SOFC Truck APU, project funded by the European Union's Seventh Framework Programme (FP7/2007–2013) for the Fuel Cells and Hydrogen

Joint Technology Initiative under grant agreement n° 278899. http://www.desta-project.eu/desta-project/. Accessed 19 Sept 2015

DIAMOND (2014) Diagnosis-aided control for SOFC power systems, project funded by the European Community's Seventh Framework Programme (FP7/2007–2013) for the Fuel Cells and Hydrogen Joint Technology Initiative under grant agreement n° 245128. http://www.diamond-sofc-project.eu. Cited 30 Aug 2015

Dolenc B, Vrečko D, Juričić Đ, Pohjoronta A, Kiviaho Pianese C (2015) Soft sensor design for estimation of SOFC stack temperatures and oxygen-to-carbon ratio. ECS Trans 68:2625–2636

Ellamla HR, Staffell I, Bujlo P, Pollet BG, Pasupathi S (2015) Current status of fuel cell based combined heat and power systems for residential sector. J Power Sources 293:312–328

Ene.field (2011) European-wide field trials for residential fuel cell micro-CHP, project funded by the European Union's Seventh Framework Programme (FP7/2007–2013) for the Fuel Cells and Hydrogen Joint Technology Initiative under grant agreement n° 303462. http://www.fch.europa.eu/project/european-wide-field-trials-residential-fuel-cell-micro-chp

Escobet T, Feroldi D, De Lira S, Puig V, Quevedo J, Riera J, Serra M (2009) Model-based fault diagnosis in PEM fuel cell systems. J Power Sources 192:216–223

Fadeyev G, Kalyakin A, Gorbova E, Brouzgou A, Demin A, Volkov A, Tsiakaras P (2015) A simple and low-cost amperometric sensor for measuring H_2, CO, and CH_4. Sens Actuators B 221:879–883

FCDISTRICT (2010) New μ-CHP network technologies for energy efficient and sustainable districts, project funded by the European Union's Seventh Framework Programme (FP7/2007–2013) for the Fuel Cells and Hydrogen Joint Technology Initiative under grant agreement n° 260105. http://www.fc-district.eu/. Accessed 19 Sept 2015

FCGEN (2011)Fuel Cell Based Power Generation, project funded by the European Union's Seventh Framework Programme (FP7/2007-2013) for the Fuel Cells and Hydrogen Joint Technology Initiative under grant agreement n° 277844. http://www.fcgen.com/. Accessed 19 Sept 2015

Ferrari ML, Traverso A, Massardo AF (2004) Transient analysis of solid oxide fuel cell hybrids: part B—anode recirculation model. In: ASME conference proceedings, pp 399–407

Gaynor R, Mueller F, Jabbari F, Brouwer J (2008) On control concepts to prevent fuel starvation in solid oxide fuel cells. J Power Sources 180(1):330–342

Gazzarri JI, Kesler O (2007) Electrochemical AC impedance model of a solid oxide fuel cell and its application to diagnosis of multiple degradation modes. J Power Sources 167:100–110

GENIUS (2009) Generic diagnosis instrument for SOFC systems, project funded by the European Community's Seventh Framework Programme (FP7/2007–2013) for the Fuel Cells and Hydrogen Joint Technology Initiative under grant agreement n° 621208. http://genius.eifer.kit.edu. Accessed 18 Aug 2015

Greco A, Sorce A, Littwin R, Costamagna P, Magistri L (2014) Reformer faults in SOFC systems: experimental andmodeling analysis, and simulated fault maps. Int J Hydrogen Energy 39:21700–21713

Guida M, Postiglione F, Pulcini G (2015) A random-effects model for long-term degradation analysis of solid oxide fuel cells. Reliab Eng Syst Safe 140:88–98

Hajimolana SA, Soroush M (2009) Dynamics and control of a tubular solid-oxide fuel cell. Ind Eng Chem Res 48:6112–6125

Holtappels P, Mehling H, Roehlich S, Liebermann SS, Stimming U (2005) SOFC system operating strategies for mobile applications. Fuel Cells 5(4):499–508

Huang Q-A, Hui R, Wang B, Zhang J (2007) A review of AC impedance model-ing and validation in SOFC diagnosis. Electrochim Acta 52:8144–8164

Huang Q-A, Wang B, Quc W, Hui R (2009) Impedance diagnosis of metal-supported SOFCs with SDC as electrolyte. J Power Sources 191:297–303

Ingimundarson A, Stefanopoulou AG, McKay A (2008) Model-based detection of hydrogen leaks in a fuel cell stack. IEEE Trans Control Syst Technol 16(5):1004–1012

Isermann R (2005) Model-based fault-detection and diagnosis—status and applications. Ann Rev Control 29:71–85

Kadowaki M (2015) Current status of National SOFC Projects in Japan. ECS Trans 68(1):15–22

Kang YW, Li J, Cao GY, Tu HY, Li J, Yang J (2009) A reduced 1D dynamic model of a planar direct internal reforming solid oxide fuel cell for system re-search. J Power Sources 188:170–176

Kneidel KE, DeBellis C, Kantak M, Norrick D, Vesely C, Palmer, BK (2004) Development of SOFC power systems using multi-layer ceramic interconnects. In: Proceedings of fuel cell seminar, November, 2004, San Antonio, Texas, USA

Larminie J, Dicks A (2003) Fuel Cell Systems Explained. John Wiley and Sons, Chichester, West Sussex, UK, pp 1–24, 207–227

Larrain D, Van Herle J, Favrat D (2006) Simulation of SOFC stack and repeat elements including interconnect degradation and anode reoxidation risk. J Power Sources 161:392–403

Leonide A, Apel Y, Ivers-Tiffee E (2009) SOFC modeling and parameter identifi-cation by means of impedance spectroscopy. ECS Trans 19:81–109

Lu N, Li Q, Sun X, Khaleel M (2006) The modeling of a standalone solid-oxide fuel cell auxiliary power unit. J Power Sources 161(2):938–948

Magistri L, Trasino F, Costamagna P (2006) Transient analysis of solid oxide fuel cell Hybrids - Part I: fuel cell models. J Eng Gas Turbines Power 128:288–293

Mahoto N, Banerjee A, Gupta A, Omar S, Balani K (2015) Progress in material selection for solid oxide fuel cell technology: a review. Prog Mater Sci 72:141–337

Marra D, Sorrentino M, Pianese C, Iwanschitz B (2013) A neural network estimator of Solid Oxide Fuel Cell performance for on-field diagnostics and prognostics applications. J Power Sources 241(2013):320–329

Marra D, Sorrentino M, Pianese C, Mennella A (2015a) A one-dimensional modelling approach for planar cylindrical solid oxide fuel cell. In: Proceedings of the ASME 2015 power and energy conversion conference, June 28–July 2, 2015, San Diego, California

Marra D, Sorrentino M, Pohjoranta A, Pianese C, Kiviaho J (2015b) A lumped dynamic modelling approach for model-based control and diagnosis of solid oxide fuel cell system with anode off-gas recycling. ECS Trans 68:3095–3106

Marwala T (2012) Condition monitoring using computational intelligence methods. Springer, London, (par 1.4)

McLarty D, Brouwer J, Samuelsen S (2013) A spatially resolved physical model for transient system analysis of high temperature fuel cells. Int J Hydrogen Energy 38:7935–7946

Milewski J, Swirski K (2009) Modelling the SOFC behaviours by artificial neural network. Int J Hydrogen Energy 34:5546–5553

Mueller F, Jabbari F, Brouwer J, Roberts R, Junker T, Ghezel-Ayagh H (2007) Control design for a bottoming solid oxide fuel cell gas turbine hybrid system. J Fuel Cell Sci Technol 4(3):221–230

Murshed AM, Huang B, Nandakumar K (2007) Control relevant modeling of planer solid oxide fuel cell system. J Power Sources 163:830–845

Nanaeda K, Mueller F, Brouwer J, Samuelsen S (2010) Dynamic modelling and evaluation of solid oxide fuel cell-combined heat and power system operating strategies. J Power Sources 195:3176–3185

Napoli R, Gandiglio M, Lanzini A, Santarelli M (2015) Techno-economic analysis of PEMFC and SOFC micro-CHP fuel cell systems for the residential sector. Energ and Buildings 103:131–146

Ormerod RM (2003) Solid oxide fuel cells, Chem Soc Rev 32:17–28

Padullés J, Ault G, McDonald J (2000) Integrated SOFC plant dynamic model for power systems simulation. J Power Sources 86(1):495–500

Papageorgopoulos D (2015) Fuel Cell Program, 2015 Annual Merit Review and Peer Evaluation Meeting, 8–12 June 2015

Patan K, (2008) Artificial neural networks for the modelling and fault diagnosis of technical process. In: Lecture notes in control and information Sciences, vol 377. Springer, p 206

Petrone R, Zheng Z, Hissel D, Péra M C, Pianese C, Sorrentino M, Becherif M, Yousfi-Steiner N (2013) A review on model-based diagnosis methodologies for PEMFCs. Int J Hydrogen Energy 38(17):7077–7091 (June)

Pollet BG, Staffell I, Shang JL (2012) Current status of hybrid, battery and fuel cell electric vehicles: from electrochemistry to market prospects. Electrochim Acta 84:235–249

Polverino P, Pianese C, Sorrentino M, Marra D (2015) Model-based development of a fault signature matrix to improve solid oxide fuel cell systems on-site diagnosis. J Power Sources 280:320–338

Polverino P, Esposito A, Pianese C, Ludwig B, Iwanschitz B, Mai A (2016) On-line experimental validation of a model-based diagnostic algorithm dedicated to a solid oxide fuel cell system. J Power Sources 306:646–657 (Accepted)

Qi Y, Huang B, Luo J (2008) 1-d dynamic modeling of SOFC with analytical so-lution for reacting gas-flow problem. AIChE J 54(6):1537–1553

Ramschak E (2009) Online PEMFC stack monitoring with "THDA". In: International symposium on diagnostic tools for fuel cell technologies, Trondheim, 23rd–24th June 2009. http://www. sintef.no/globalassets/project/fc-tools/dokumenter/presentation/6a/ramschak.pdf, Cited 7 Sept 2015

Riascos LAM, Simoes MG, Miyagi PE (2008) On-line fault diagnostic system for proton exchange membrane fuel cells. J Power Sources 175:419–429

Roehrens D, Han F, Haydn M, Schafbauer W, Sebold D, Menzler NH, Buchkremer HP (2015) Advances beyond traditional SOFC cell designs. Int J Hydrogen Energy 40:11538–11542

SAFARI (2014) SOFC APU for auxiliary road-truck installations, project funded by the European Union's Seventh Framework Programme (FP7/2007–2013) for the Fuel Cells and Hydrogen Joint Technology Initiative under grant agreement n° 325323. http://www.adelan.co.uk/ projects-2/examples-of-live-projects/safari/. Accessed 19 Sept 2015

Sandhu GS, Rattan KS (1997) Design of a neuro-fuzzy controller, Systems, Man, and Cybernetics, 1997. In: 1997 IEEE international conference on computational cybernetics and simulation, vol 4, pp 3170–3175, 12–15 Oct 1997

Sedghisigarchi K, Feliachi A (2004a) Dynamic and transient analysis of power distribution systems with fuel Cells-part I: fuel-cell dynamic model. IEEE Trans Energy Convers 19:423–428

Sedghisigarchi K, Feliachi A (2004b) Dynamic and transient analysis of power distribution systems with fuel Cells-part II: control and stability enhancement. IEEE Trans Energy Convers 19:429–434

Selimovic A, Kemm M, Torisson T, Assadi M (2005) Steady state and transient thermal stress analysis in planar solid oxide fuel cells. J Power Sources 145:463–469

Singhal SC (2002) Solid oxide fuel cells for stationary, mobile and military applications. Solid State Ionics 152–153:405–410

Singhal S C, Kendall K (2003) High-temperature solid oxide fuel cells: fundamentals, design and applications. First edn. Elsevier Science (ed)

Sohlberg B (2003) Grey box modelling for model predictive control of a heating process. J Process Control 13:225–238

Sorce A, Greco A, Magistri L, Costamagna P (2014) FDI oriented modeling of an experimental SOFC system, model validation and simulation of faulty states. Appl Energy 136:894–908

Sorrentino M, Pianese C (2009a) Control oriented modeling of solid oxide fuel cell auxiliary power unit for transportation applications. ASME Trans J Fuel Cell Sci Technol 6:041011–04101112

Sorrentino M, Pianese C (2009b) Grey-Box modeling of SOFC unit for design, control and diagnostics applications. European Fuel Cell Forum, June 29–July 2009, Lucerne, Switzerland

Sorrentino M, Pianese C (2011) Model-based development of low-level control strategies for transient operation of solid oxide fuel cell systems. J Power Sources 196:9036–9045

Sorrentino M, Pianese C, Guezennec YG (2008) A hierarchical modeling ap-proach to the simulation and control of planar solid oxide fuel cells. J Power Sources 180:380–392

Sorrentino M, Marra D, Pianese C, Guida M, Postiglione F, Wang K, Pohjoranta A (2014) On the use of neural networks and statistical tools for nonlinear modeling and on-field diagnosis of solid oxide fuel cell stacks. Energy Procedia 45:298–307

Tsikonis L, Albrektsson J, Van herle J, Favrat D (2014) The effect of bias in gas temperature measurements on the control of a Solid Oxide Fuel Cells system. J Power Sources 245:19–26

Vandersteen, JDJ, Kenney B, Pharoah, JG Karan K (2004) Mathematical modelling of the transport phenomena and the chemical/electrochemical reactions in solid oxide fuel cells: a review. In: Proceedings of Canadian hydrogen and fuel cells conference, Toronto (Canada)

Virkar AV (2007) A model for solid oxide fuel cell (SOFC) stack degradation. J Power Sources 172:713–724

Vrečko D, Dolanc G, Dolenc B, Vrančić D, Pregelj B, Marra D, Sorrentino M, Pianese C, Pohjoranta A, Juričić Đ (2015) Feedforward-feedback control of a SOFC power system: a simulation study. ECS Trans 68:3151–3163

Wahl S, Segarra A, Horstmann P, Carré M, Bessler WG, Lapicque F, Friedrich KA (2015) Modeling of a thermally integrated 10 kWe planar solid oxide fuel cell system with anode offgas recycling and internal reforming by discretiza-tion in flow direction. J Power Sources 279:656–666

Wang K, Hissel D, Péra MC, Steiner NY, Marra D, Sorrentino M, Pianese C, Monteverde M, Cardone P, Saarinen J (2011) A Review on solid oxide fuel cell models. Int J Hydrogen Energy 36(12):7212–7228

Wang K, Péra MC, Hissel D, Steiner NY, Pohjoranta A, Pofahl S (2012) SOFC modelling based on discrete Bayesian network for system diagnosis use. IFAC Proc Volumes 8:675–680

Weber A, Tiffée EI (2004) Materials and concepts for solid oxide fuel cells (SOFCs) in stationary and mobile applications. J Power Sources 127:273–283

White BM, Lundberg WL, Pierre JF (2015) Accomplishments, Status, and Roadmap for the U.S. Department of Energy's Fossil Energy SOFC Program. ECS Trans 68(1):23–38

Wu XJ, Zhu XJ (2011) Multi-loop control strategy of a solid oxide fuel cell and micro gas turbine hybrid system. J Power Sources 196:8444–8449

Wu XJ, Huang W, Zhu XJ (2011) Thermal modeling of a solid oxide fuel cell and micro gas turbine hybrid power system based on modified LS-SVM. Int J Hydrogen Energy 36:885–892

Xi H, Sun J, Tsourapas V (2007) A control oriented low order dynamic model for planar SOFC using minimum Gibbs free energy method. J Power Sources 165(1):253–266

Yu SK (2015) New and renewable energy in Korea—best practices in policy and development. In: IPHE—23rd Steering Committee meeting, Wuhan, China, 27-28 May 2015

Zhang L, Li X, Jiang J, Li S, Yang J, Li J (2015) Dynamic modeling and analysis of a 5-kW solid oxide fuel cell system from the perspectives of cooperative control of thermal safety and high efficiency. Int J Hydrogen Energy 40:456–476

Zheng Z, Petrone R, Péra M C, Hissel D., Becherif M., Pianese C, Yousfi-Steiner N, Sorrentino M (2013) A review on non-model based diagnosis methodologies for PEM fuel cell stacks and systems. Int J Hydrogen Energy 38(21):8914–8926 (July)

Zhu Y, Tomsovic K (2002) Development of models for analyzing the load-following performance of microturbines and fuel cells. Electr Power Syst Res 62:1–11

Chapter 2
Models Hierarchy

2.1 From Physical to Synthesis Models

Nowadays, engineering research always relies on the mathematical representation of the physical system under-investigation to ensure proper model-based optimization, control, and diagnosis task be performed, aiming at continuously improving performance, while still achieving cost-effectiveness targets.

Physical (i.e., white-box) models are obtained by suitably adapting well-known physical principles and knowledge to a plant or a process, in such a way as to define the governing equations of a specific phenomenon (Romijn et al. 2008). On the other hand, black-box models correspond to a purely mathematical description of the phenomenon that must be completely inferred from a suitable experimental data set (i.e., identification or training data set) and tested for generalization verification on further experimental data sets (i.e., test sets). Therefore, black-box models do not rely on physical content, thus resulting in lower computational burden since, as an example, no differential equations, neither ordinary nor partial, are to be solved. On the other hand, their accuracy necessarily relies on extended experimental campaigns. Regressions and neural networks are typical examples of black-box models, also known as synthesis models, since their mathematical representation is flexible enough to be applied for mapping any continuous functional relationship between outputs and inputs of a given process with a very low computational burden as compared to physical (i.e., white-box) models.

Between white- and black-box, gray-box models fall (Romijn et al. 2008). Such models are usually developed exploiting both physical content and experimental data availability, thus resulting into a middle level model category that exhibits a good compromise between computational and experimental burden. Typical examples of gray-box models are zero-dimensional or lumped models, in which a few or just one variable are assumed as state variables to characterize system behavior both in steady-state and transient conditions (Ramallo-González et al. 2013). Figure 2.1 shows the qualitative variation of required experimental burden as

© Springer-Verlag London 2016
D. Marra et al., *Models for Solid Oxide Fuel Cell Systems*,
Green Energy and Technology, DOI 10.1007/978-1-4471-5658-1_2

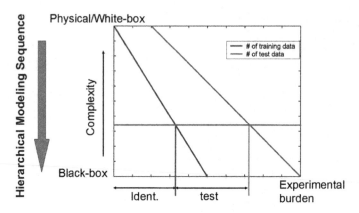

Fig. 2.1 Qualitative description of the impact of modeling approach on required experimental burden

a function of physical content. It is reasonable to expect that physical models require relatively few experimental data for their validation. On the other hand, black-box approaches entail performing a high number of experiments to be used for both identification and test.

The most suitable modeling approach has to be selected depending upon the specific application field, as shown on Table 2.1. As expected, high physical content is required to improve component design. Particularly in the fuel cell field, it is known that the main cell manufacturers utilize highly complex physical models to improve cell geometries and select the most performing materials (US Department of Energy 2004).

Regarding system sizing and optimal control strategies definition, black-box and gray-box models are both more suitable than physical, high computational intensive models, especially when destined to real-time tasks. Optimal balance of plant can be achieved adopting large-scale design optimization algorithms, which usually require several function evaluations (Rizzoni et al. 2005; Arsie et al. 1999). Therefore, the use of models with a good compromise between accuracy and computational burden is highly beneficial.

Table 2.1 Modeling approaches versus application fields

Modeling approach	Application fields		
	Component design	System sizing	Optimal control and diagnosis
Physical	X		
Steady gray/black-box		X	X
Dynamic gray/black-box		Possible	X
Hybrid modeling	Not applicable	Appropriate combination of different approaches	

Optimal control strategies are defined at both supervisory (i.e., high control level) and low control level (Arsie et al. 2010; Guezennec et al. 2003). The definition of the optimal working set-points, which competes to higher levels, does not necessarily require dynamic simulations. On the other hand, low-level control tasks are often accomplished via feedback strategies, thus requiring to take into account the main system dynamics (Pukrushpan et al. 2004). In both cases, suitable optimization analyses are required to define the best strategies, thus suggesting the combined use of steady and dynamic gray-/black-box models. Moreover, many plants, such as SOFC-based auxiliary power units (APUs), consist of devices with time constants that significantly differ from each other. Thus, proper combination of dynamic and steady gray-/black-box submodels (i.e., hybrid modeling) might allow to further reduce computational time.

Finally, model-based monitoring, feedback control, as well as model-based diagnosis, entail extending model applicability to real-time tasks, thus outlining once again the need of adopting hybrid approaches, preferably encouraging the use of very low computational intensive models, such as gray- and especially black-box ones.

The joint analysis of Fig. 2.1 and Table 2.1 indicates a significant conflicting need, which has to be carefully accounted for in the trade-off analysis on modeling approach: experimental burden. Such an issue can be addressed by recurring to a hierarchical approach. Particularly, the use of physical models can be optimized, in that once tested for validation with a reduced amount of experimental data, they can be used not only for single component optimal (see Table 2.1) but also as virtual-experiments generators. In such a way, the available reference data sets can be extended and then, following the hierarchical sequence, both gray-and black-box models can be identified and validated without further impact on experimental burden. Models hierarchy is of course also suitable to individuate the best balance between more and less physical models into a hybrid modeling architecture (see Table 2.1). In the following section, the hierarchical sequence proposed by the authors to suitably address model-based design, control, and diagnosis of SOFC systems is presented and discussed in detail.

2.1.1   Hierarchical Approach to SOFC Modeling

The hierarchical approach proposed in this book is illustrated in Fig. 2.2. The levels 1, 2, and 3 are representative of different physical contents. At the highest level (i.e., 1) is the real system, which of course may correspond either to an SOFC stack or to a fully operational SOFC system. Depending on which experimental set-up is currently available, the number of SOFC model-based development tasks, which might be conducted mainly exploiting real observations/measurements, will vary, thus indirectly but still effectively impacting on how to find the most suitable model development path.

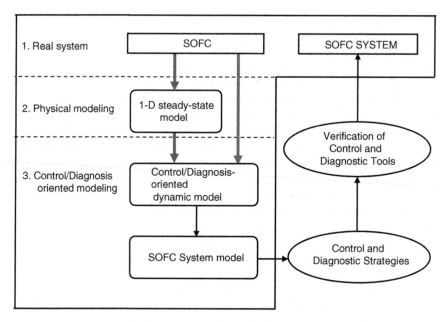

Fig. 2.2 Hierarchical approach for real-time modeling and model-based control and diagnosis of SOFC systems

Particularly, Fig. 2.2 indicates that two different paths (i.e., blue and red lines) can be followed to develop control- and diagnosis-oriented models (i.e., level 3 models). The former, which usually corresponds to limited availability of experiments or, as mentioned above, to less informative data sets as single SOFC stack characterization might be, first entails developing more physical models (i.e., level 2) prior to switch to level 3, where more computationally effective models are built. Such an articulated path (i.e., the blue one) is justified by the need for ensuring sufficiently extended data sets be available for identifying and validating less physical models, as shown on Fig. 2.2. On the other hand, whenever more experiments can be run on the real system, without reducing cost-effectiveness of the experimental campaign, the latter red path can be chosen. If this is the case, either gray-box or black-box or even hybrid models can be directly developed, i.e., without developing level 2 models, thus positively impacting on the development burden of advanced model-based control and diagnosis tools, which are the expected output of the hierarchical procedure shown on Fig. 2.2. Such a figure also highlights the importance of accounting for the opportunity of getting back to the physical plant for final validation of the developed model-based tools.

The next section deepens the methodological aspects associated to the development of the models hierarchy, starting from the description of SOFC one-dimensional modeling and then proceeding to treating synthesis models definition and development, according to the above discussed hierarchical approach.

2.2 Dimensional Modeling

The adoption of one-dimensional (1D) modeling approach for planar SOFCs allows meeting significantly conflicting needs, such as good accuracy and affordable computational burden. On one hand, the literature (Chan et al. 2001; Aguiar et al. 2005) clearly indicates how the correct estimation of current and temperature spatial distributions in the flow direction enhances the evaluation of effective cell performance. On the other hand, removing the need for solving the governing equations in the other dimensions contributes to greatly reducing computational time. Therefore, 1D modeling approach emerges as a very promising way to address SOFC-related optimization/design problems, such as balance of plant analyses and polarization models identification (Braun 2002).

Furthermore, the appreciable physical content characterizing 1D modeling ensures satisfactory flexibility to extend model validity to different cell geometries, materials, and fuels (fully or partially reformed petroleum fuels). The above positive features are particularly useful to provide "virtual" experimental data sets that suitably cover the typical SOFC operating domain. The latter aspect serves at accomplishing one of the main targets of this book, namely the exploitation of models hierarchy to develop real-time applicable model-based control and diagnostic tools.

2.2.1 The 1D Model

The proposed 1D model aims at describing spatial evolution of current, temperature and gas composition within a rectangular-planar SOFC, both in co- and counterflow configuration. Therefore, it consists of an integrated structure of submodels, mass, energy, and electrochemical balance, which are run in an iterative procedure. Particularly, spatial variations in the cell are modeled by discretizing the domain into computational elements along cell length. At each element, mass, energy, and electrochemical balances are solved.

The following hypotheses were assumed to model both hydrogen and methane reformate fed solid oxide fuel cells (Ormerod 2003; Braun 2002; Sorrentino et al. 2008):

- isopotential interconnect and electrodes (Braun 2002);
- completely stirred conditions are considered at the element level;
- air and fuel fed gases are treated separately as perfect gases;
- pressure drop within fuel and air channels is neglected, according to the indications provided in Burt et al. (2004);
- the flow velocity is constant over the channel cross section;
- the concentration of reactants is constant over the channel cross section;
- diffusion of reactants along the flow channel is negligible as convection prevails on transport;

- cell boundaries are adiabatic;
- radiative heat transfer between solid trilayer and metallic interconnects is assumed negligible.

In the following subsections, the basic working principle of SOFC is initially recalled, prior to describing with deep details the above-mentioned mass, energy, and electrochemical balances.

2.2.1.1 Mass Balance

Applying conservation of mass to a generic control volume Ω surrounding the entire cell shown in Fig. 2.3, the steady-state material balance for each specie m can be expressed as:

$$\int_{\Omega} \vec{\nabla} \cdot \vec{N}_m \, dV = 0 \quad m = [H_2, CH_4, H_2O, CO, CO_2, N_2, O_2] \tag{2.1}$$

Anode and cathode are discretized in N computational element in the flow direction, as shown on Fig. 2.3, where inlet, outlet and source (or sink) molar flows are represented for the ith computational element and the m_{an}th and m_{ca}th specie.

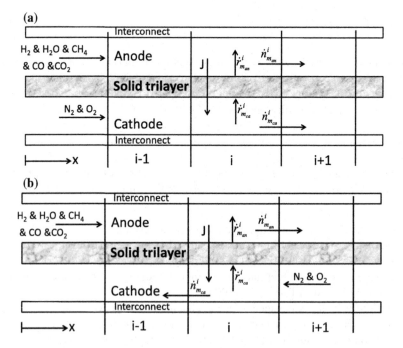

Fig. 2.3 Mass balance discretization for co-flow (**a**) and counter-flow (**b**) planar solid oxide fuel cell

Applying Eq. (2.1) to the discretized cell for all the species, the following mass balances are obtained for co- (see Fig. 2.3a) and counterflow configurations (see Fig. 2.3b):

$$0 = \dot{n}_{H_2}^{i-1} - \dot{n}_{H_2}^{i} + 3 \cdot \dot{r}_{ref}^{i} + \dot{r}_{shift}^{i} - \dot{r}_{ox}^{i} \tag{2.2a}$$

$$0 = \dot{n}_{H_2O}^{i-1} - \dot{n}_{H_2O}^{i} - \dot{r}_{ref}^{i} - \dot{r}_{shift}^{i} + \dot{r}_{ox}^{i} \tag{2.2b}$$

$$0 = \dot{n}_{CH_4}^{i-1} - \dot{n}_{CH_4}^{i} - \dot{r}_{ref}^{i} \tag{2.2c}$$

$$0 = \dot{n}_{CO}^{i-1} - \dot{n}_{CO}^{i} + \dot{r}_{ref}^{i} - \dot{r}_{shift}^{i} \tag{2.2d}$$

$$0 = \dot{n}_{CO_2}^{i-1} - \dot{n}_{CO_2}^{i} + \dot{r}_{shift}^{i} \tag{2.2e}$$

$$\begin{cases} 0 = \dot{n}_{N_2}^{i-1} - \dot{n}_{N_2}^{i}, & \text{co-flow} \\ 0 = \dot{n}_{N_2}^{i+1} - \dot{n}_{N_2}^{i}, & \text{counter-flow} \end{cases} \tag{2.2f}$$

$$\begin{cases} 0 = \dot{n}_{O_2}^{i-1} - \dot{n}_{O_2}^{i} - 0.5 \cdot \dot{r}_{ox}^{i}, & \text{co-flow} \\ 0 = \dot{n}_{O_2}^{i+1} - \dot{n}_{O_2}^{i} - 0.5 \cdot \dot{r}_{ox}^{i}, & \text{counter-flow} \end{cases} \tag{2.2g}$$

The quantities $\dot{r}_{ref}^{i}$, $\dot{r}_{shift}^{i}$ and $\dot{r}_{ox}^{i}$ appearing in Eq. (2.2a)–(2.2g) are the reaction rates (mol s^{-1}) of the methane reforming (see Eq. 2.3a), water–gas shift (see Eq. 2.3b) and electro-oxidation (see Eq. 2.3c) reaction, respectively:

$$CH_4 + H_2O \rightarrow CO + 3H_2 \tag{2.3a}$$

$$CO + H_2O \leftrightarrow CO_2 + H_2 \tag{2.3b}$$

$$H_2 + \frac{1}{2}O_2 \rightarrow H_2O \tag{2.3c}$$

The water gas shift reaction (Eq. 2.3b) is assumed to be in equilibrium, with equilibrium constant expressed as:

$$K_{shift}^{i}(p^i, T_s^i) = \exp\left(\frac{-\Delta G_{shift}(p^i, T_s^i)}{R \cdot T_s^i}\right) \tag{2.4}$$

K_{shift} can also be estimated as a function of molar fractions (Massardo and Lubelli 2000):

$$K_{shift}^{i} = \frac{x_{H_2}^{i} \cdot x_{CO_2}^{i}}{x_{CO}^{i} \cdot x_{H_2O}^{i}} = \frac{\left(\dfrac{\dot{n}_{H_2}^{i-1} + 3\dot{r}_{ref}^{i} + \dot{r}_{shift}^{i} - \dot{r}_{ox}^{i}}{\dot{n}_{tot}^{i-1} + 2\dot{r}_{ref}^{i}}\right) \cdot \left(\dfrac{\dot{n}_{CO_2}^{i-1} + \dot{r}_{shift}^{i}}{\dot{n}_{tot}^{i-1} + 2\dot{r}_{ref}^{i}}\right)}{\left(\dfrac{\dot{n}_{CO}^{i-1} + \dot{r}_{ref}^{i} - \dot{r}_{shift}^{i}}{\dot{n}_{tot}^{i-1} + 2\dot{r}_{ref}^{i}}\right) \cdot \left(\dfrac{\dot{n}_{H_2O}^{i-1} - \dot{r}_{ref}^{i} - \dot{r}_{shift}^{i} + \dot{r}_{ox}^{i}}{\dot{n}_{tot}^{i-1} + 2\dot{r}_{ref}^{i}}\right)} \tag{2.5}$$

The reaction rates $\dot{r}_{ox}^i$ and $\dot{r}_{ref}^i$ (see Eq. 2.2a–2.2g) are computed via Faraday's law (see Eq. 2.6) and the temperature and pressure dependent correlation (see Eq. 2.7) indicated by Achenbach and Riensche (1994), respectively:

$$\dot{r}_{ox}^i = \frac{J^i \cdot A^i}{F \cdot n_e} \tag{2.6}$$

$$\dot{r}_{ref}^i = 4274 \cdot p_{CH_4}^i \cdot A^i e^{-(82000/R/T_s^i)} \tag{2.7}$$

The last reaction rate $\dot{r}_{shift}^i$ is determined solving the system of Eqs. (2.4)–(2.7).

Boundary conditions of the systems of Eq. (2.2a)–(2.2g) are the inlet flows, estimated as a function of fuel utilization (Eq. 2.8) and excess air (Eq. 2.9a, 2.9b):

$$U_f = \begin{cases} \dfrac{\dot{n}_{H_2,react}}{\dot{n}_{H_2}^0} = \dfrac{\sum_{i=1}^N \dot{r}_{ox}^i}{\dot{n}_{H_2}^0} = \dfrac{\sum_{i=1}^N J^i \cdot A^i}{F \cdot n_e} \cdot \dfrac{1}{\dot{n}_{H_2}^0} & \text{pure } H_2 \text{ feed} \\[4mm] \dfrac{\dot{n}_{H_2,react}}{4\dot{n}_{CH_4}^{pre,in}} = \dfrac{\sum_{i=1}^N \dot{r}_{ox}^i}{4\dot{n}_{CH_4}^{pre,in}} = \dfrac{\sum_{i=1}^N J^i \cdot A^i}{F \cdot n_e} \cdot \dfrac{1}{4\dot{n}_{CH_4}^{pre,in}} & \text{reformate feed} \end{cases} \tag{2.8}$$

$$\lambda = \frac{\dot{n}_{O_2}^0}{\dot{n}_{O_2,sto}} = \frac{\dot{n}_{O_2}^0}{\frac{1}{2}\dot{n}_{H_2,react}} = \frac{\dot{n}_{air}^0}{4.76 \cdot \frac{1}{2}\dot{n}_{H_2,react}} \quad \text{co-flow} \tag{2.9a}$$

$$\lambda = \frac{\dot{n}_{O_2}^{N+1}}{\dot{n}_{O_2,sto}} = \frac{\dot{n}_{O_2}^{N+1}}{\frac{1}{2}\dot{n}_{H_2,react}} = \frac{\dot{n}_{air}^{N+1}}{4.76 \cdot \frac{1}{2}\dot{n}_{H_2,react}} \quad \text{counter-flow} \tag{2.9b}$$

It is worth noting that the denominator of Eq. (2.9a), (2.9b) for reformate feed equals 4 times the methane flow at pre-reformer inlet. This holds because, for each CH_4 molecule, four H_2 molecules can be obtained, as indicated by Eq. (2.3a)–(2.3c). The use of a pre-reforming stage, to partially pre-reform methane, is required to avoid the endothermic internal reforming reaction (see Eq. 2.3a) to cause excessive temperature drop at anode inlet. Moreover, it is evident that while Eq. (2.9a), (2.9b) allows directly evaluating the incoming air flows, to be fed as input to the system of Eqs. (2.2a)–(2.2g) and (2.8) must be suitably post-processed, in case of reformate feed, to precisely estimate the anode inlet flows. The following subsection treats in detail the numerical approach to be followed to perform such an estimation.

Estimation of Anode Inlet Flows

As shown in Fig. 2.4, an external pre-reformer is usually placed upstream the anode inlet. Therefore, the anode composition at anode inlet (i.e., pre-reformer outlet) will depend on the following parameters: pre-reformer conversion factor (i.e., REF) and SOFC inlet (i.e., pre-reformer outlet, see Fig. 2.4) temperature and pressure.

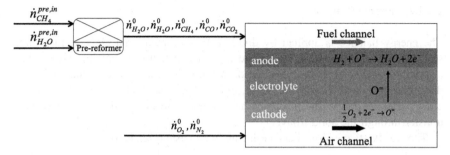

Fig. 2.4 Schematic representation of main variables involved in the procedure used to estimate anode inlet fuel composition (see Eqs. 2.10, 2.11a, 2.11b, 2.11c and 2.12)

Pre-reformer conversion factor is defined as the ratio between the methane flow entering the SOFC and the methane flow at pre-reformer inlet:

$$\text{REF} = \frac{\dot{n}^0_{CH_4}}{\dot{n}^{pre,in}_{CH_4}} \tag{2.10}$$

where $\dot{n}^{pre,in}_{CH_4}$ and $\dot{n}^0_{CH_4}$ are the methane flow entering the pre-reformer and the SOFC stack, respectively. Assuming a steam-pre-reformer is used, water has also to be fed, as shown on Fig. 2.4. Since in the pre-reformer both reforming (Eq. 2.3a) and water-gas-shift reaction (Eq. 2.3b) take place, the following balances hold for carbon, hydrogen, and oxygen, respectively:

$$\dot{n}^0_{CH_4} + \dot{n}^0_{CO} + \dot{n}^0_{CO_2} - \dot{n}^{pre,in}_{CH_4} = 0 \tag{2.11a}$$

$$4\dot{n}^0_{CH_4} + 2\dot{n}^0_{H_2} + 2\dot{n}^0_{H_2O} - 4\dot{n}^{pre,in}_{CH_4} - 2\dot{n}^{pre,in}_{H_2O} = 0 \tag{2.11b}$$

$$\dot{n}^0_{H_2O} + \dot{n}^0_{CO} + 2\dot{n}^0_{CO_2} - \dot{n}^{pre,in}_{H_2O} = 0 \tag{2.11c}$$

In order to determine the five unknowns $\dot{n}^0_{H_2}$, $\dot{n}^0_{H_2O}$, $\dot{n}^0_{CH_4}$, $\dot{n}^0_{CO}$ and $\dot{n}^0_{CO_2}$ as a function of reformer conversion factor and anode inlet temperature and pressure, the system of Eqs. (2.10) and (2.11a, 2.11b, 2.11c) must be coupled to a fifth equation (i.e., Eq. 2.12), which is directly derived from the water-gas-shift reaction equilibrium constant (see Eq. 2.5). Such a fifth equation, holding valid at the SOFC inlet (i.e., pre-reformer outlet), can be simply expressed as:

$$K_{shift}\left(p^0, T^0_{an}\right) = \frac{\dot{n}^0_{H_2} \cdot \dot{n}^0_{CO_2}}{\dot{n}^0_{H_2O} \cdot \dot{n}^0_{CO}} \tag{2.12}$$

where K_{shift} is evaluated by means of Eq. (2.4).

2.2.1.2 Energy Balance

The energy balance is applied by dividing the computational element into three separate control volumes, namely solid trilayer and fuel (i.e., anode) and air (i.e., cathode) channels (see Fig. 2.5). The dominant effects described in the model are the convective heat transfer between solid trilayer and fuel and air streams, the energy transfer due to the reactants, and products flows and the conductive heat transfer through the solid trilayer.

The steady-state energy balance for an open system reads as:

$$\frac{dE}{dt} = \dot{E}_{in} - \dot{E}_{out} + \dot{Q} - \dot{W} = 0 \tag{2.13}$$

In the following subsections, the way Eq. (2.13) was particularized to each control volume is presented and discussed, along with basic theoretical principles through which each energy flow term shown on Fig. 2.5 was treated.

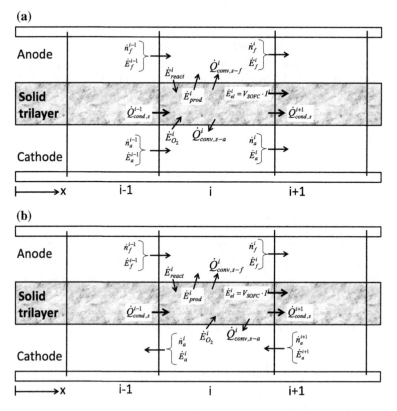

Fig. 2.5 Energy balance discretization for co-flow (**a**) and counter-flow (**b**) planar solid oxide fuel cell

Energy Balance for the Fuel Channel

As stated above, the energy balance for the fuel channel was obtained by assuming the only heat exchange mechanisms to correspond to: (i) the energy rates associated with species flow in and out of the ith computational element; (ii) convective heat exchange with the solid trilayer and (iii) the energy rates released by the reactions occurring at the anode surface, i.e., reforming, water-gas-shift and electro-oxidation. On the other hand, heat exchange with the surrounding environment is neglected in accordance to the adiabatic cell hypothesis introduced above. Therefore, since no useful work is delivered, the energy balance for the fuel channel can be written in the following compact form:

$$0 = \dot{E}_{f,\text{in}} - \dot{E}_{f,\text{out}} + \dot{Q}_{\text{conv},s-f} \tag{2.14}$$

where the energy rates entering and leaving the ith computational element (see Fig. 2.4) are evaluated as follows:

$$\dot{E}_{f,\text{in}} = \dot{E}_f^{i-1} + \dot{E}_{\text{prod}}^i \tag{2.15}$$

$$\dot{E}_{f,\text{out}} = \dot{E}_f^i + \dot{E}_{\text{react}}^i \tag{2.16}$$

Particularly, the energy rates associated with inlet and outlet flows and the electro-oxidation, reforming, and water-gas-shift reactions, are calculated as follows (Braun 2002):

$$\begin{aligned} \dot{E}_f^{i,i-1} = {} & \dot{n}_{\text{H}_2}^{i,i-1} \cdot h_{\text{H}_2}\left(T_f^{i,i-1}\right) + \dot{n}_{\text{CH}_4}^{i,i-1} \cdot h_{\text{CH}_4}\left(T_f^{i,i-1}\right) \\ & + \dot{n}_{\text{H}_2\text{O}}^{i,i-1} \cdot h_{\text{H}_2\text{O}}\left(T_f^{i,i-1}\right) + \dot{n}_{\text{CO}_2}^{i,i-1} \cdot h_{\text{CO}_2}\left(T_f^{i,i-1}\right) \\ & + \dot{n}_{\text{CO}}^{i,i-1} \cdot h_{\text{CO}}\left(T_f^{i,i-1}\right) \end{aligned} \tag{2.17}$$

$$\begin{aligned} \dot{E}_{\text{prod}}^i = {} & h_{\text{H}_2\text{O}}\left(T_s^i\right) \cdot \dot{r}_{\text{ox}}^i + \left[3 \cdot h_{\text{H}_2}\left(T_s^i\right) + h_{\text{CO}}\left(T_s^i\right)\right] \cdot \dot{r}_{\text{CH}_4}^i \\ & + \left[h_{\text{H}_2}\left(T_s^i\right) + h_{\text{CO}_2}\left(T_s^i\right)\right] \cdot \dot{r}_{\text{shift}}^i \end{aligned} \tag{2.18}$$

$$\begin{aligned} \dot{E}_{\text{react}}^i = {} & h_{\text{H}_2}\left(T_f^i\right) \cdot \dot{r}_{\text{ox}}^i + \left[h_{\text{CH}_4}\left(T_f^i\right) + h_{\text{H}_2\text{O}}\left(T_f^i\right)\right] \cdot \dot{r}_{\text{CH}_4}^i \\ & + \left[h_{\text{CO}}\left(T_f^i\right) + h_{\text{H}_2\text{O}}\left(T_f^i\right)\right] \cdot \dot{r}_{\text{shift}}^i \end{aligned} \tag{2.19}$$

On the other hand, the convective heat transfer between solid trilayer and fuel channel (see Fig. 2.5) is evaluated as:

$$\dot{Q}_{conv,s-f} = \bar{h}_f^i \cdot A_s^i \cdot \left(T_s^i - T_f^i\right) \tag{2.20}$$

where $A_s^i = w_{ch} \cdot l_{ch} \cdot N^{-1}$ is the heat transfer area of the ith computational element. The convective heat transfer coefficient $\bar{h}_f^i$ is estimated, at each slice, as a function of fuel thermal conductivity (k_f), which in turn depends on both fuel composition and temperature:

$$\bar{h}_f^i = \frac{N_u \cdot k_f^i}{D_h} \tag{2.21}$$

$$k_f^i = \sum_{j=1}^{n} \frac{x_j^i \cdot k_j\left(T_f^i\right)}{\sum_{m=1}^{n} k_m\left(T_f^i\right) \cdot z_{jm}} \tag{2.22}$$

where the quantity z_{jm} is estimated as a function of the thermal conductivity and molar mass of both jth and mth species:

$$z_{jm} = \frac{\left[1 + (k_j/k_m)^{1/2} \cdot (M_m/M_j)^{1/4}\right]^2}{\left[8(1 + M_j/M_m)\right]^{1/2}} \tag{2.23}$$

Equation (2.21) also highlights the dependence of heat transfer coefficient on hydraulic diameter (Iwata et al. 2000) and Nusselt number (Braun 2002), here evaluated through the following relationships:

$$D_h = \frac{4A_{ch}}{2(h_{ch} + w_{ch})} = \frac{2w_{ch} \cdot h_{ch}}{(h_{ch} + w_{ch})} \tag{2.24}$$

$$\begin{aligned} Nu = 7.541 \cdot 1 - (2.61 \cdot \beta + 4.97 \cdot \beta^2 - 5.119 \cdot \beta^3 \\ + 2.702 \cdot \beta^4 + 0.548 \cdot \beta^5) \end{aligned} \tag{2.25}$$

where β is the ratio between channel height (h_{ch}) and channel width (w_{ch}).

Energy Balance for the Air Channel

In this case, the energy balance was written in such a way as to include, beyond the species flows entering and leaving the ith computational element, also the oxygen mass flow through the solid electrolyte (see Fig. 2.5). Thus, the following energy balance results for the air channel:

$$0 = \dot{E}_{a,\text{in}} - \dot{E}_{a,\text{out}} + \dot{Q}_{\text{conv},s-a} \qquad (2.26)$$

$$\dot{E}_{a,\text{in}} = \dot{E}_a^{i-1} \quad \text{co-flow} \qquad (2.27a)$$

$$\dot{E}_{a,\text{in}} = \dot{E}_a^{i+1} \quad \text{counter-flow} \qquad (2.27b)$$

$$\dot{E}_{a,\text{out}} = \dot{E}_a^i + \dot{E}_{O_2}^i \qquad (2.28)$$

where the energy rates associated with inlet and outlet flows are estimated as follows:

$$\dot{E}_a^{i,i-1,i+1} = \dot{n}_{O_2}^{i,\,i-1,i+1} \cdot h_{O_2}\left(T_a^{i,i-1,i+1}\right) + \dot{n}_{N_2} \cdot h_{N_2}\left(T_a^{i,i-1,i+1}\right) \qquad (2.29)$$

$$\dot{E}_{O_2}^i = \frac{\dot{r}_{\text{ox}}^i}{2} h_{O_2}\left(T_a^i\right) \qquad (2.30)$$

Finally, the heat exchange between air channel and solid trilayer ($\dot{Q}_{\text{conv},s-a}$) was modeled extending the approach outlined from Eqs. (2.20)–(2.25) to the species flowing in the air channel (i.e., oxygen and nitrogen).

Energy Balance for the Solid Trilayer

The last energy balance involves the control volume surrounding the entire solid trilayer and the interconnect (i.e., electrodes, electrolyte, and interconnect). Provided that radiative heat transfer can be safely neglected, the inclusion of the interconnect enables correct evaluation of heat transfer through the solid without the need of adding two further energy balances, i.e., fuel side- and air side-interconnect, as suggested by Braun (2002). Therefore, the energy rates to be included in the balance are related to the reactions occurring at electrodes surface, the heat exchange with fuel and air channels and the conductive heat transfer across the solid itself. Therefore, the following equation was obtained:

$$0 = \dot{E}_{\text{react}}^i - \dot{E}_{\text{prod}}^i + \dot{E}_{O_2}^i - \dot{Q}_{\text{conv},s-f}^i - \dot{Q}_{\text{conv},s-a}^i - \dot{E}_{el}^i + \Delta\dot{Q}_{\text{cond},s}^i \qquad (2.31)$$

Among the seven terms appearing at the right-hand side of Eq. (2.31), only two of them were not introduced yet, namely the useful work released to the external load, evaluated via Eq. (2.32), and the conductive heat transfer. The last term was estimated (Braun 2002) as the difference between the heat coming from the $i - 1$th slice and the heat leaving the ith computational element and heading for the $i + 1$th slice (see Fig. 2.5 and Eq. 2.33):

$$\dot{E}^i_{el} = V_{SOFC} \cdot J^i \cdot A^i \tag{2.32}$$

$$\Delta \dot{Q}^i_{cond,s} = \frac{k_s}{\Delta X} \cdot A_{front} \left(T^{i-1}_s - T^i_s \right) - \frac{k_s}{\Delta X} \cdot A_{front} \left(T^i_s - T^{i+1}_s \right) \tag{2.33}$$

where ΔX equals the entire cell/channel length (l_{ch}) divided by the number of computational elements (**N**). The term V_{SOFC} at the right-hand side of Eq. (2.32) is evaluated through the electrochemical model described in Sect. 2.2.1.3. On the other hand, the term k_s at the right-hand side of Eq. (2.33) represents the solid thermal conductivity, which is usually higher in case of anode electrodes with a high percentage of Nickel cermets (Larrain et al. 2003). Furthermore, A_{front} is the solid frontal area, which is constant along the flow direction for rectangular planar SOFC designs. A_{front} is calculated as a function of electrodes, electrolyte, and interconnect thickness, thus lumping together the effects played by conduction heat transfer through solid trilayer and interconnect.

2.2.1.3   Electrochemical Balance

There are three major forms of polarization losses in fuel cells: activation, ohmic, and concentration losses. In some cases, a minor constant offset also contributes to the total polarization, which is the result of minor losses, such as internal current and leaks (e.g., fuel crossover, Chick et al. 2003). The sum of the different polarizations results in the voltage drop from Nernst potential (see Eq. 2.35) to effective operating value (0., V_{SOFC}, see Eq. 2.34). Since interconnect and electrodes are assumed isopotential, cell voltage is constant over the whole cell, and thus, the electrochemical balance can be written as follows:

$$0 = E^i_{Nernst} - V^i_{Act} - V^i_{Ohm} - V^i_{Conc} - V_{SOFC} \tag{2.34}$$

where the Nernst potential is estimated as:

$$E^i_{Nernst} = -\frac{\Delta G^i_{ox}(T^i_s)}{n_e F} - \frac{R_u T^i_s}{n_e F} \ln \left(\frac{p^i_{H_2O}}{p^i_{H_2} \sqrt{p^i_{O_2}}} \right) \tag{2.35}$$

Upon knowledge of SOFC voltage, the total power drawn out of a single cell can be evaluated as:

$$\dot{E}_{el} = V_{SOFC} \cdot I_{SOFC} = V_{SOFC} \cdot \sum_{i=1}^{N} \left(J^i \cdot A^i \right) \tag{2.36}$$

In the general fuel cell literature, the three main polarization losses (i.e., activation, ohmic, and concentration) are widely modeled by means of Eq. 2.37 [known as Tafel equation (Noren and Hoffman 2005)] through Eq. 2.39 (Braun 2002):

$$V_{Act}^i = \frac{R \cdot T_s^i}{\alpha(T_s^i) \cdot F} \cdot \sinh^{-1}\left(\frac{J^i}{2 J_0(T_s^i)}\right) \tag{2.37}$$

$$V_{ohm,k}^i = \frac{l_k}{\sigma_k(T_s^i)} \cdot J^i$$

$$V_{ohm}^i = \sum_k V_{ohm,k}^i \quad k = [an, ca, el] \tag{2.38}$$

$$V_{Conc}^i = -\frac{RT_s^i}{2F} \cdot \left[\frac{1}{2}\ln\left(1 - \frac{J^i}{J_{cs}}\right) + \ln\left(1 - \frac{J^i}{J_{as}}\right) - \ln\left(1 + \frac{p_{H_2}^i \cdot J^i}{p_{H_2O}^i \cdot J_{as}}\right)\right] \tag{2.39}$$

Figure 2.6 shows the variation of activation losses as a function of current density and operating temperature. Activation polarization increases with a higher slope at low current densities (especially at low temperatures, as shown on Fig. 2.6) as compared to higher values, thus confirming that such a polarization is dominant at low loads (Larminie and Dicks 2003).

Figure 2.7 illustrates the effect of current density and temperature on SOFC ohmic losses. Even in this case temperature increase has a significantly positive effect on cell performance. Furthermore, Fig. 2.7 highlights the importance of operating fuel cells mostly at medium loads, mainly to avoid excessive ohmic resistance to occur. Using anode-supported (see Fig. 2.7b) instead of electrolyte-supported (Fig. 2.7a) cell structure allows reducing, such an overpotential, due to the higher conductivity of both electrodes as compared to electrolyte (see Table 2.2).

Fig. 2.6 Activation losses dependence on current and temperature. The values shown in the figure were estimated via Eq. (2.37). Further details on the parameters utilized in Eq. (2.37) can be retrieved from Sorrentino et al. (2004)

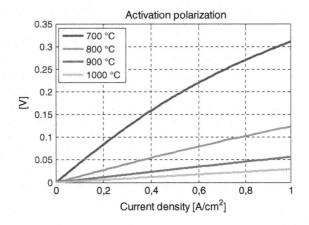

Fig. 2.7 Ohmic losses dependence on current and temperature for electrolyte-(**a**) and anode-supported (**b**) SOFC (see cell specifications in Table 2.2). The values shown in the figure were estimated via Eq. (2.38). Further details on the parameters utilized in Eq. (2.38) can be retrieved from Sorrentino et al. (2004)

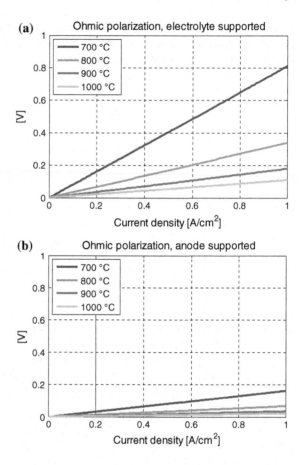

Figure 2.8 shows concentration polarization variation as a function of current density and temperature. Since both anode and cathode limiting current densities (see Eq. 2.39) depend on species diffusivity (Chick et al. 2003; Yang et al. 2013), the values shown in Fig. 2.8 were calculated assuming spatially averaged concentrations for the species involved in reformate-fed SOFC functioning (see Fig. 2.12). Concentration losses slightly increase with temperature, thus exhibiting an opposite behavior with respect to both activation and ohmic polarization (Yang et al. 2013). In any case, the comparative analysis of Figs. 2.6, 2.7, and 2.8 indicates that the negative temperature effect on concentration overpotential plays a very minor effect, thus not impacting on the general provision to make SOFC systems operate at the highest possible temperature. Furthermore, concentration polarization for anode-supported cell (see Fig. 2.8b) was shown to be higher than electrolyte-supported (see Fig. 2.8a), in accordance with the results presented and discussed in (Chan et al. 2001). Nevertheless, nowadays many stack manufacturers prefer adopting electrode-supported cells, due to the strong reduction in ohmic

Table 2.2 Input data and main cell specifications of the two model benchmarks (Ferguson et al. 1996; GENIUS 2013), both referring to a rectangular-planar SOFC

Cell type	Electrolyte-supported				Anode-supported
Validation benchmark	ES_1	ES_2	ES_3	ES_4	AS
Electroactive area (A) (cm^2)	100				100
Cell length (l_{ch}) (cm)	10				10
Channel width (w_{ch}) mm)	3				1
Rib width (w_{rib}) (mm)	2.42				1
Channel height (h_{ch}) (mm)	1				1
Number of channels	18				50
Anode thickness (l_{an}) (μm)	50				1000
Cathode thickness (l_{ca}) (μm)	50				50
Electrolyte thickness (l_{el}) (μm)	150				30
Interconnect thickness (l_{int}) (μm)	2500				1100
Pressure (bar)	1				1
Inlet temperature (T^0) (°C)	900				[700 750]
Fuel utilization (U_f)	0.85				[0.5 0.6]
Excess air (λ)	7				5
Average current density (J) (A cm^{-2})	0.3				[0.3 0.35]
ASR formula (Ω cm^2)	$ASR^i = \frac{l_{an}}{\sigma^i_{an}} + \frac{l_{ca}}{\sigma^i_{ca}} + \frac{l_{int}}{\sigma^i_{int}} + 3\frac{l_{el}}{\sigma^i_{el}}$				$10^{-4} \cdot \exp\left[\frac{5 \cdot 10^4}{R} \cdot \left(\frac{1}{T^i_s} - \frac{1}{973}\right)\right]$
Anode conductivity (S cm^{-1})	$\sigma^i_{an} = \frac{9.5 \cdot 10^7}{T^i_s} \cdot \exp\left(-\frac{1150}{T^i_s}\right)$				–
Cathode conductivity (S cm^{-1})	$\sigma^i_{ca} = \frac{4.2 \cdot 10^7}{T^i_s} \cdot \exp\left(-\frac{1200}{T^i_s}\right)$				–
Electrolyte conductivity (S cm^{-1})	$\sigma^i_{el} = 3.34 \cdot 10^4 \cdot \exp\left(-\frac{10300}{T^i_s}\right)$				–
Interconnect conductivity (S cm^{-1})	$\sigma^i_{int} = \frac{9.3 \cdot 10^6}{T^i_s} \cdot \exp\left(-\frac{1100}{T^i_s}\right)$				–

(continued)

Table 2.2 (continued)

Cell type	Electrolyte-supported				Anode-supported
Thermal conductivity (k_s) (W m^{-1} K^{-1})	2				20
Fuel feed	Pure H_2	Reformate fuel	Pure H_2	Reformate fuel	Reformate fuel
Cell flow configuration	Co-flow		Counter-flow		Counter-flow
H_2O/CH_4 fraction $\left(\dot{n}_{H_2O}^{pre,in} / \dot{n}_{CH_4}^{pre,in} \right)$	–	2.5	–	2.5	2
Inlet H_2 composition	0.9	0.263	0.9	0.263	0.128
Inlet H_2O composition	0.1	0.493	0.1	0.493	0.560
Inlet CH_4 composition	–	0.171	–	0.171	0.28
Inlet CO composition	–	0.029	–	0.029	0.0004
Inlet CO_2 composition	–	0.044	–	0.044	0.032
Inlet O_2 composition	0.21				
Inlet N_2 composition	0.79				

losses and the very minor contribution of concentration losses on overall SOFC overpotential (Aguiar et al. 2004; Larrain et al. 2003).

It is finally worth remarking at the end of this section that, since SOFCs are mainly operated in the linear current-voltage profile region, some researchers (Andersson 2011) proposed to lump the total amount of polarization losses into one contribution, proportional to the area-specific resistance coefficient (ASR), thus resulting in the following simplified electrochemical balance:

$$0 = E_{Nernst}^i - ASR^i \cdot J^i - V_{SOFC} \qquad (2.40)$$

Standing the high reliability of Eq. (2.40), in this work the 1D model validation was performed by referring to ASR formulas proposed in the literature.

2.2.1.4 Load Balance

The last balance, to be satisfied to close the system of one-dimensional governing equations, is obtained by imposing that average current density along the cell channels corresponds to the applied load. This assumption well agrees with the

Fig. 2.8 Concentration losses dependence on current and temperature for electrolyte-(**a**) and anode-supported (**b**) SOFC (see cell specifications in Table 2.2). The values shown in the figure were estimated via Eq. (2.39). Further details on the parameters utilized in Eq. (2.39) can be retrieved from Sorrentino et al. (2004)

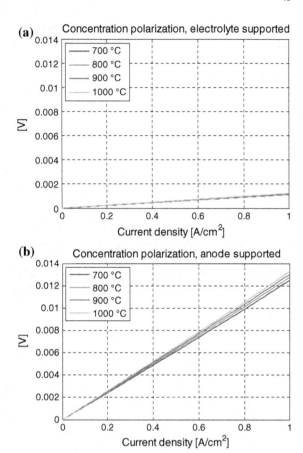

most common operation strategy adopted for fuel cells, namely to change current density to meet useful power demand, which in turn corresponds to galvanostatic operation of the stack. The mathematical counterpart of the above concept can be represented as:

$$0 = \frac{I_{\text{SOFC}}}{A} - \frac{1}{N}\sum_{i=1}^{N} J^i \qquad (2.41)$$

where the term at the first denominator corresponds to the entire electroactive area of a single fuel cell.

2.2.1.5   Numerical Solution

The spatial variation of the main operating variables is evaluated by discretizing the SOFC in N computational elements and iteratively solving, for each element, the governing equations. Particularly, energy, electrochemical, and current density balances are to be solved to estimate the distributions of current density, temperatures, and compositions along the cell channels. The numerical problem can thus be expressed as:

$$F(\Theta) = 0 \qquad (2.42)$$

where Θ is a $5N + 1$ long vector of unknowns:

$$\Theta = \left\{ \theta^1, \theta^2, \ldots, \theta^N, V_{SOFC} \right\}$$
$$\theta^i = \left\{ J^i, r^i_{shift}, T^i_f, T^i_a, T^i_s \right\} \qquad (2.43)$$

As explained in the previous sections, the system of governing equations is highly nonlinear, thus requiring to build-up a numerical iterative solution (Braun 2002; Haynes 1999; Sorrentino et al. 2004). Such a solution is here obtained as shown in Fig. 2.9, which highlights how the problem expressed by Eq. (2.42) is treated as a multi-objective optimization task, mathematically expressed as:

$$\min_{\Theta} |F(\Theta)| \qquad (2.44)$$

Equation (2.44) ensures that the optimal Θ vector, obtained after successfully minimizing all of the f^i terms shown in Eq. (2.51), is as close as possible to those satisfying Eq. (2.42) exactly (Yang 2013). The vector F is calculated at each h iteration by concatenating the residuals associated to the electrochemical balance (i.e., res_{ele}, Eq. 2.45a, 2.45b), water-gas-shift equilibrium constant (i.e., res_{Kshift}, Eq. 2.46), energy balances (i.e., $res_{energy,f}$, Eq. 2.47, $res_{energy,a}$, Eq. 2.48, $res_{energy,s}$, Eq. 2.49), and load balance (i.e., res_{load}, Eq. 2.50):

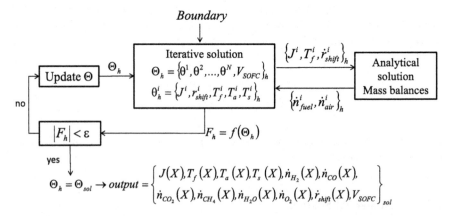

Fig. 2.9 Schematic representation of the procedure developed to solve, in a least square way, the system of governing equations expressed by Eqs. (2.45a, 2.45b)–(2.50). The variables $\dot{n}^i_{fuel}$ and $\dot{n}^i_{air}$ are nothing but vectors including the molar flows of anodic and cathodic species, respectively

$$\text{res}_{\text{ele}}^i = E_{\text{Nernst}}^i - V_{\text{ohm}}^i - V_{\text{conc}}^i - V_{\text{act}}^i - V_{\text{SOFC}} \tag{2.45a}$$

$$\text{res}_{\text{ele}}^i = E_{\text{Nernst}}^i - \text{ASR}^i \cdot J^i - V_{\text{SOFC}} \tag{2.45b}$$

$$\text{res}_{K\text{shift}}^i = \exp\left(\frac{-\Delta G_{\text{shift}}\left(T_s^i\right)}{R \cdot T_s^i}\right) - \frac{\dot{n}_{H_2}^i \cdot \dot{n}_{CO_2}^i}{\dot{n}_{CO}^i \cdot \dot{n}_{H_2O}^i} \tag{2.46}$$

$$\text{res}_{\text{energy},f}^i = \dot{E}_{f,\text{in}} - \dot{E}_{f,\text{out}} + \dot{Q}_{\text{conv},s-f} \tag{2.47}$$

$$\text{res}_{\text{energy},a}^i = \dot{E}_{a,\text{in}} - \dot{E}_{a,\text{out}} + \dot{Q}_{\text{conv},s-a} \tag{2.48}$$

$$\text{res}_{\text{energy},s}^i = \dot{E}_{\text{react}}^i - \dot{E}_{\text{prod}}^i + \dot{E}_{O_2}^i - \dot{Q}_{\text{conv},s-f}^i - \dot{Q}_{\text{conv},s-a}^i - \dot{E}_{\text{el}}^i + \Delta \dot{Q}_{\text{cond},s}^i \tag{2.49}$$

$$\text{res}_{\text{load}} = \frac{I_{\text{SOFC}}}{A} - \frac{1}{N}\sum_{i=1}^{N} J^i \tag{2.50}$$

Thus, the following 5N + 1 long vector F results at each h iteration:

$$\begin{aligned} F &= \left\{f^1, f^2, \ldots, f^N, \text{res}_{\text{load}}\right\} \\ f^i &= \left\{\text{res}_{\text{ele}}^i, \text{res}_{K\text{shift}}^i, \text{res}_{\text{energy},f}^i, \text{res}_{\text{energy},a}^i, \text{res}_{\text{energy},s}^i\right\} \end{aligned} \tag{2.51}$$

It is worth noting here that the residuals (i.e., the res terms), introduced in Eqs. 2.45a, 2.45b–2.50, are nothing but the right-hand side of their respective governing equations, presented in the previous sections (e.g., res_{load} corresponds to the value of the right-hand side of Eq. 2.41).

Figure 2.9 highlights how the residual balances to be met, namely the mass balances expressed by Eq. 2.2a–2.2g, are not directly involved in the multi-objective minimization problem described above, in that no molar flows are fed as unknown parameters to the problem expressed by Eq. (2.44). This happens both because the water-gas-shift reaction rate is one of the unknowns (see Eq. 2.43) and, furthermore, the methane reforming and electro-oxidation reaction rates are respectively computed as a function of solid temperature (see Eq. 2.6) and current density (see Eq. 2.7), which are also present among the unknowns. Therefore, as shown in Fig. 2.9, at each iteration, beyond calculating the actual value of the vector F, the mass balances expressed by Eq. 2.2a–2.2g are solved analytically, thus yielding on output the molar flows distributions. Such molar distributions are then concatenated to the vector of unknowns (i.e., Eq. 2.43), thus resulting in the following comprehensive output vector yielded by the entire procedure schematically described in Fig. 2.9:

$$\text{Output} = \left\{ \begin{array}{l} J(X), T_f(X), T_a(X), T_s(X), \dot{n}_{\text{H}_2}(X), \dot{n}_{\text{CO}}(X), \\ \dot{n}_{\text{CO}_2}(X), \dot{n}_{\text{CH}_4}(X), \dot{n}_{\text{H}_2\text{O}}(X), \dot{n}_{\text{O}_2}(X), \dot{r}_{\text{shift}}(X), V_{\text{SOFC}} \end{array} \right\} \tag{2.52}$$

Figure 2.9 also highlights that the iterative procedure, to be developed to perform root-finding for the problem expressed by Eq. (2.44), can only be started upon knowledge of the vector of boundary conditions, which includes the following variables:

$$\text{Boundary} = \left\{ U_f, \lambda, \overline{J}, T^0, p, x^0_{f,m}, x^0_{a,m} \right\} \tag{2.53}$$

Finally, it is worth shortly describing how the procedure illustrated in Fig. 2.9 operates. Once the boundary conditions expressed by Eq. (2.53) are known, the iterative procedure is started by providing as inputs the initial 5N + 1 conditions for the vector of unknowns Θ to the multi-objective optimization problem, here solved by means of the well-known Levenberg–Marquardt minimization algorithm (Yang 2013). Then, at each h iteration the f^i_h values are determined as a function of the current Θ vector, depending on which the molar flow distributions also are analytically computed with (see "Analytical solution" block in Fig. 2.9). Afterwards, the vector F is checked in a least square way, namely if the norm $|F|$ is below a safe threshold, the actual values of the variables included in Eq. (2.43) are yielded on output as final solution of the system of governing equations. Otherwise, the procedure is repeated by updating the Θ vector by means of the Levenberg–Marquardt algorithm, until a final solution with a satisfactorily small norm $|F|$ is reached.

2.2.1.6 Model Validation

The model detailed above was tested for validation by referring to two different benchmarks. The first one (i.e., benchmark ES in Table 2.2, corresponding to a high temperature electrolyte-supported SOFC), which was already tested by the authors in a previous article (Sorrentino et al. 2008) for co-flow configuration only, was here further tested in counterflow configuration, thus resulting in four sub-benchmarks ES1 through ES4, as explained in Table 2.2. The second benchmark (i.e., benchmark AS in Table 2.2, corresponding to a low temperature anode-supported SOFC) was tested by authors' institution, i.e., the University of Salerno (UNISA, Italy), the Technical Institute of Finland (VTT), and the University of Genoa (UNIGE, Italy), within the European Funded project GENIUS (GENIUS 2013). Both model benchmarks were solved assuming a simplified ASR formula, thus allowing to perform lumped modeling of polarization losses as a function of operating temperature (see Eq. 2.45b). Figure 2.10 provides the reader with a schematic explanation of the main geometrical data introduced in Table 2.2.

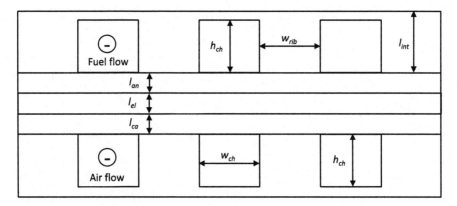

Fig. 2.10 Cross-sectional view of a single, planar, and co-flow SOFC consisting of three rectangular channels at both fuel and air side

The following two subsections present a deep discussion on the main outcomes of the above-introduced model benchmarking, performed by discretizing the cell into a number of computational elements N ranging between 10 and 50, depending on successful convergence of the root-finding problem shown in Fig. 2.9. As a consequence of different cell discretization, running time varied between 20 and 400 s on a 2.53 GHz Intel® Core™ Duo CPU. Afterwards, a parametric analysis is proposed in Sect. 2.2.1.7, whose aim is to show the potentialities, offered by the 1D model, to qualitatively analyze the effects played by main operating variables on SOFC performance and behavioral features.

ES Benchmark Results

The benchmarks ES_1 through ES_4 were modeled assuming that cell polarization losses can be estimated via the ASR formula proposed by the International Energy Agency (IEA) (Braun 2002; Ferguson 1996), which is also given in Table 2.2. The results obtained running the 1D SOFC model on the ES cases are summarized in Figs. 2.11, 2.12 and Table 2.3. It is worth noting here that in Fig. 2.11 only the solid-trilayer temperature profile is plotted, since the temperature of the other control volumes (i.e., fuel and air channels) do not differ significantly, as already observed in previous studies (Iwata et al. 2000).

In case of pure hydrogen feed (see subplots (a) and (c) in Figs. 2.11 and 2.12), the model results indicate that temperature monotonically increases in the direction of air-flow. Such a behavior is expected since SOFCs are air-cooled by supplying the cathode channels with extremely high excess of air with respect to stoichiometry. Therefore, the air-flow is subject to continuous heat adsorption going from

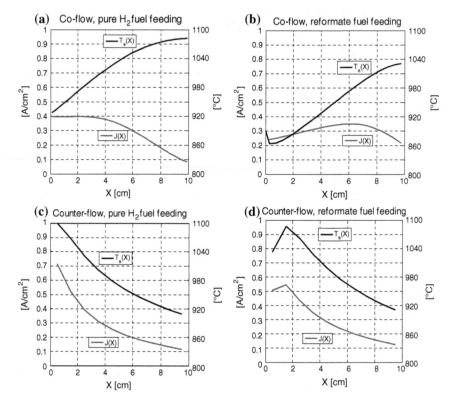

Fig. 2.11 Current and temperature distributions yielded on output by the 1D model executed on BA benchmarks (see Table 2.2)

cathode inlet to outlet. On the other hand, when reformate fuel is fed to the anode channels (see subplots (b) and (d) in Figs. 2.11 and 2.12), temperature distribution exhibits a sudden decrease in proximity of anode inlet. This happens because the residual methane, held by the inlet fuel flow after having undergone the external pre-reforming process (see Sect. 2.2.1.2.1), quickly gets internally reformed, as shown in Fig. 2.12b, d. Such an internal reforming process is endothermic, thus causing solid temperature to decrease until methane disappears, around one fifth of cell length. Afterwards, temperature trend is similar to that occurring in case of pure hydrogen feed.

The analysis of current distributions is shown in Figs. 2.11 and 2.12 highlights that they are strongly dependent on both temperature and hydrogen composition. Indeed, current always decreases when hydrogen composition becomes very low. However, in case of co-flow configuration the current profile is more flat as

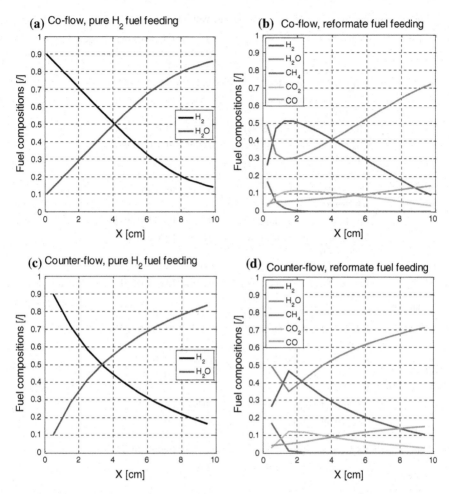

Fig. 2.12 Fuel compositions distributions yielded on output by the 1D model executed on the ES benchmarks (see Table 2.2)

compared to counterflow cases. Such a behavior can be justified considering that, as explained in the previous paragraph, counterflow temperature distribution reaches the maximum values nearby anode inlet, whereas the opposite happens in co-flow cases. As it is well known (Braun 2002; U.S. Department of Energy 2004), temperature increase always causes polarization losses to decrease. Therefore, counterflow current profiles start (at $X = 0$) from very high values at anode inlet, due to the beneficial effects of temperature increase and high hydrogen concentration. On the other hand, in co-flow configuration current profile starts with a flat shape in

Table 2.3 Comparison of 1D SOFC model results with literature data

Variable	Benchmark	1D SOFC model	Literature data ranges (min max) (Braun 2002; Ferguson et al. 1996)
Voltage (V)	ES$_1$ (co-flow, pure H$_2$)	0.713	[0.702 0.722]
	ES$_2$ (co-flow, methane reformate)	0.656	[0.633 0.650]
	ES$_3$ (counter-flow, pure H$_2$)	0.715	[0.709 0.730]
	ES$_4$ (counter-flow, methane reformate)	0.693	[0.680 0.693]
Maximum current density (A cm^2)	ES$_1$ (co-flow, pure H$_2$)	0.397	[0.373 0.396]
	ES$_2$ (co-flow, methane reformate)	0.350	[0.3040 0.367]
	ES$_3$ (counter-flow, pure H$_2$)	0.715	[0.711 0.897]
	ES$_4$ (counter-flow, methane reformate)	0.544	[0.533 0.655]
Minimum current density (A cm^2)	ES$_1$ (co-flow, pure H$_2$)	0.085	[0.102 0.137]
	ES$_2$ (co-flow, methane reformate)	0.218	[0.175 0.251]
	ES$_3$ (counter-flow, pure H$_2$)	0.116	[0.108 0.124]
	ES$_4$ (counter-flow, methane reformate)	0.125	[0.099 0.133]
Maximum solid temperature (°C)	ES$_1$ (co-flow, pure H$_2$)	1082	[1048 1098]
	ES$_2$ (co-flow, methane reformate)	1031	[1020 1034]
	ES$_3$ (counter-flow, pure H$_2$)	1099	[1062 1084]
	ES$_4$ (counter-flow, methane reformate)	1087	[1062 1089]
Minimum solid temperature (°C)	ES$_1$ (co-flow, pure H$_2$)	927	[924 930]
	ES$_2$ (co-flow, methane reformate)	864	[845 862]
	ES$_3$ (counter-flow, pure H$_2$)	909	[904 913]
	ES$_4$ (counter-flow, methane reformate)	911	[906 915]

case of pure hydrogen feeding, while, when reformate fuel is used, current is even lower than maximum values at anode inlet. Lower currents at anode inlet, in case of co-flow configuration, are determined by the high polarization losses caused by low temperature values in those locations. The difference between pure hydrogen and reformate fuel feeding is due again to the reforming process occurring nearby cathode inlet, which causes temperature to be further low at those locations. It is of course straightforward that minimum current values occur near anode outlet, since hydrogen gets almost completely depleted and, thus, concentration overpotential induced by fuel starvation becomes dominant.

Regarding fuel compositions shown in Fig. 2.12, one significant difference emerges from the comparative analysis of co- and counterflow configurations, namely the different slope with which hydrogen composition decreases. Indeed, counterflow hydrogen composition decreases with a larger slope at the beginning, as it can be easily noted by looking at the much earlier intersection occurring among H_2 and H_2O distributions as compared to co-flow configuration. Particularly, in case of pure hydrogen feeding the intersection corresponds to the ordinate value of 0.5, which is reached after 4 and 3.2 cm for co- (Fig. 2.12a) and counterflow (Fig. 2.12c), respectively. On the other hand, in case of reformate fuel, the intersection of interest is the second one, since the first one occurs because hydrogen/water concentration initially increases/decreases due to the reforming reaction, which mostly occurs nearby the anode inlet. In this case, the intersection corresponds to an ordinate value of 0.4, which is reached by hydrogen and water concentrations after 4 and 2.2 cm in case of co- (Fig. 2.12b) and counterflow (Fig. 2.12d), respectively. This behavior is strictly linked to what was explained in the previous paragraphs about temperature and current. Particularly, the higher the temperature the lower polarization losses will be, thus the electrochemical reaction mostly occurs nearby anode inlet in counterflow, which in turn results in faster H_2 depletion at those channel locations.

The reliability of the 1D SOFC model here presented and, as a consequence, the validity of the considerations made above, is confirmed by the very good agreement emerging from the results comparison described in Table 2.3. Particularly, 1D SOFC model outputs always fall within or very close to the data ranges provided in the literature for the IEA benchmark. Furthermore, the comparison between pure hydrogen and reformate fuel voltages indicates that the latter feeding always results in lower SOFC voltage, due to the lower average temperature as compared to the hydrogen feed (see Fig. 2.11). This does not mean that adopting a reformate fuel results in less efficient SOFC operation, due to the different lower heating values of the working fuels, methane, and hydrogen, respectively. Indeed, efficiencies in the two cases are evaluated as follows:

$$\eta_{H_2} = \frac{V_{SOFC} \cdot \bar{J} \cdot A}{n_{H_2}^0 \cdot LHV_{H_2}} \tag{2.54}$$

$$\eta_{CH_4} = \frac{V_{SOFC} \cdot \bar{J} \cdot A}{n_{CH_4}^{pre,in} \cdot LHV_{CH_4}} \tag{2.55}$$

Substituting Eq. (2.8) in Eqs. (2.54) and (2.55), the efficiencies in cases ES_1 and ES_3 (i.e., pure hydrogen, see Table 2.2) and cases ES_2 and ES_4 (i.e., reformate fuel) are, respectively:

$$\eta_{H_2} = \frac{V_{SOFC} \cdot \bar{J} \cdot A}{\frac{\bar{J} \cdot A}{2F \cdot U_f} \cdot LHV_{H_2}} = \frac{2F}{LHV_{H_2}} \cdot V_{SOFC} \cdot U_f = 0.8 \cdot 0.713 \cdot 0.85$$

$$= 0.485 \, \text{co-flow} = 0.8 \cdot 0.715 \cdot 0.85 = 0.486 \, \text{counter-flow}$$

(2.56)

$$\eta_{CH_4} = \frac{V_{SOFC} \cdot \bar{J} \cdot A}{\frac{\bar{J} \cdot A}{8F \cdot U_f} \cdot LHV_{CH_4}} = \frac{8F}{LHV_{CH_4}} \cdot V_{SOFC} \cdot U_f = 0.96 \cdot 0.656 \cdot 0.85$$

$$= 0.535 \, \text{co-flow} = 0.96 \cdot 0.693 \cdot 0.85 = 0.565 \, \text{counter-flow}$$

(2.57)

It is worth noting that Eqs. (2.56) and (2.57) enable fast and easy estimation of SOFC gross efficiency, which is mainly dependent on voltage (easily measurable on real systems) and fuel utilization values.

AS Benchmark Results

As mentioned above, the second benchmark was specifically created, within the EU project GENIUS (2013), aiming at performing a comparative analysis between variables estimated via three different 1D static models. The results of such a comparative analysis, shown in Figs. 2.13, 2.14 and 2.15, were considered worthy to be included in this book, not only to further assess the reliability of the 1D model proposed by the authors, but also to examine the performance of a planar SOFC with geometry and operating temperatures significantly different than ES benchmark (see Table 2.2). Particularly, the anode-supported cell (i.e., the one modeled in benchmark AS) is quite representative of what is currently addressed by stack manufacturers as the most suitable configuration to meet the conflicting requirements of lifetime, performance and cost, with the final aim of enhancing SOFC commercial deployment (Blum et al. 2013; Topsoe Fuel Cell 2013; Sofcpower 2013).

Figure 2.13 highlights the very good agreement achieved by the involved institutions[1] on SOFC voltage estimation, with values ranges bounded within 8.8 mV. On the other hand, Figs. 2.14 and 2.15 allow appreciating the fact that all models predicted similar temperature and current spatial distributions, also with similar shapes, especially when comparing VTT and UNISA data.

The comparative analysis of current and temperature profiles, predicted in the ES (specifically ES_4) and AS benchmarks (i.e., Fig. 2.11 with Figs. 2.14 and 2.15), clearly outlines that, despite being both operated in counterflow, the AS profiles are much smoother than ES_4. This can be explained considering that in the AS benchmark a higher thermal conductivity was assumed with respect to ES_4 (see Table 2.2). Such an assumption, which could represent a case in which a higher percentage of Nickel cermets is present in the anode material, makes the effect of

---

[1]UNISA model corresponds to the 1D SOFC model label of Table 2.3.

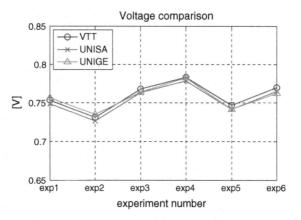

Fig. 2.13 Comparison of single cell voltages estimated by the three 1D SOFC models on the AS benchmark. Exp 1 through exp 6 corresponds to the specification provided in the plot titles shown on Fig. 2.14 (Reprinted with permission of original authors —GENIUS 2013)

thermal conduction through the solid (see Eq. 2.33) more relevant, which in turn results in the more even temperature distributions shown on Fig. 2.15, as compared to Fig. 2.11d. As deeply discussed in Sect. 2.2.1.6.1, the flatter temperature profile of Fig. 2.15 causes also the current profile to smooth out (see Fig. 2.14), due to strict dependence of current density on temperature through polarization losses.

2.2.1.7 Steady-State Parametric Analysis

This section presents an analysis and discusses related outcomes on the effects played by main operating variables, namely fuel utilization, excess air, and temperature, on SOFC performance in steady conditions. For this purpose, the cell geometry and polarization losses model used for the AS benchmark (see Table 2.2) were referred to when performing a dedicated 1-D model-based numerical analysis.

The analysis was carried-out for both pure H_2 and methane reformate feeds, by separately varying the operating variables around the nominal values listed in Table 2.4. Hereinafter, atmospheric operation is assumed (i.e., operating pressure was set to 1 bar). It is finally worth mentioning that only results obtained in co-flow configuration are shown hereinafter for sake of conciseness, as a consequence of the fact that the impact played by main operating variables on cell performance is substantially the same as in the counterflow configuration.

Figure 2.16 shows the cause-effect diagram associated to temperature increase. At high current loads, cell voltage increases, due to the beneficial impact of temperature rise on both activation and ohmic polarization (Braun 2002; Haynes 1999). On the other hand, at low loads the reduction in Nernst potential slightly prevails on losses decrease. The above cause-effect relationships are consistent with the parametric analysis carried-out on temperature influence, as shown in Fig. 2.17. Temperature increase throughout the cell length was simply obtained by increasing inlet temperature. The comparison between the Fig. 2.17a, b highlights that the

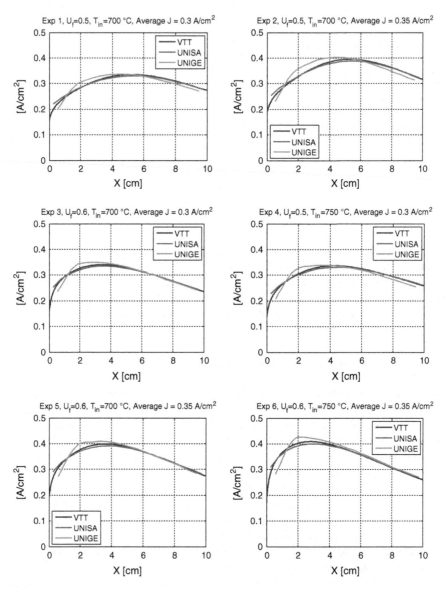

Fig. 2.14 Comparison of current density profiles yielded on output by the three models involved in the AS benchmark (Reprinted with permission of original authors—GENIUS 2013)

slope of the voltage curves obtained for methane reformate feed is higher than pure H_2 feed.

Figure 2.18 shows the cause-effect diagram associated to fuel utilization variation. Fuel utilization influence on cell voltage is twofold. Whenever fuel utilization

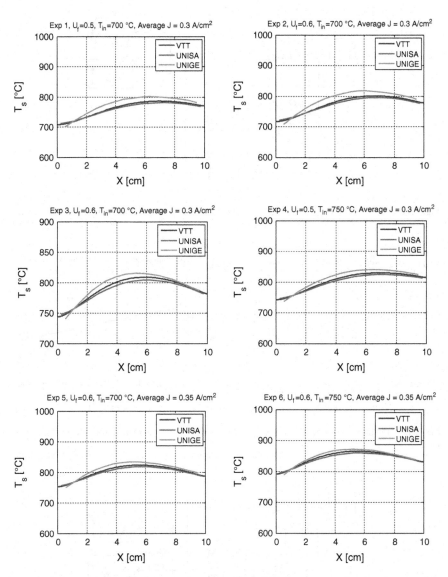

Fig. 2.15 Comparison of solid temperature profiles yielded on output by the three models involved in the AS benchmark (Reprinted with permission of original authors—GENIUS 2013)

Table 2.4 Nominal operating conditions assumed in the parametric analysis

Parameter	Value
U_f (/)	0.65
λ (/)	7
T_{in} (°C)	750

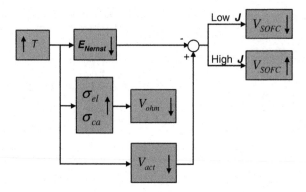

Fig. 2.16 Cause-effect diagram showing the overall effect of temperature increase on cell voltage

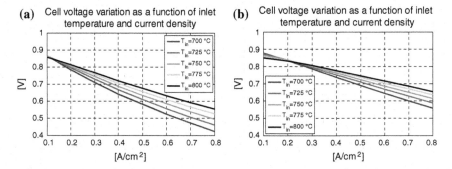

Fig. 2.17 Impact of temperature increase on SOFC voltage for **a** methane reformate and **b** pure H_2 feed

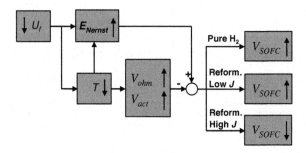

Fig. 2.18 Cause-effect diagram showing the overall effect of fuel utilization variation on cell voltage

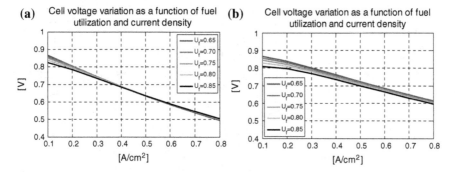

Fig. 2.19 Impact of fuel utilization variation on SOFC voltage for **a** methane reformate and **b** pure H_2 feed

reduces, reactant partial pressure in the fuel channel increases, thus resulting in higher Nernst potential throughout the cell. On the other hand, a higher amount of fuel results in temperature decrease because of two reasons: more heat adsorbed by flowing gases and, for reformate feed, more heat required for internally reforming the residual amount of methane. In Fig. 2.19 the results obtained when varying fuel utilization are shown. For pure H_2 feed (see Fig. 2.19b), a voltage gain is observed throughout the investigated load range of current, although it decreases when current density increase. In case of methane reformate feed, instead, U_f effect is positive at low loads, then slightly negative at medium-high current densities (see Fig. 2.19a). Therefore, the indications provided in the cause-effect diagram were shown to be consistent.

In Fig. 2.20, the cause-effect diagram describing the excess air effect is shown. Such an operating variable, similarly to what was discussed previously for the temperature effect, influences cell performance in two ways. A high excess of air positively affects Nernst potential, due to the higher reactants partial pressure at the cathode side. On the other hand, a high amount of air overcools the cell, thus causing activation and ohmic polarization to increase. The dominance of one effect

Fig. 2.20 Cause-effect diagram showing the overall effect of excess air variation on cell voltage

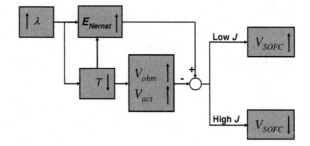

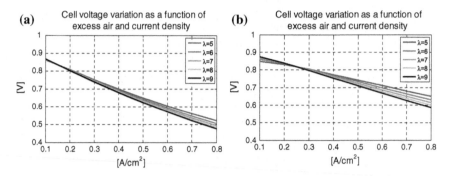

Fig. 2.21 Impact of excess air variation on SOFC voltage for **a** methane reformate and **b** pure H_2 feed

on the other results in the occurrence of two separate effect zones, as confirmed by the parametric analysis performed on λ, whose results are shown on Fig. 2.21. The comparison between methane reformate (see Fig. 2.21a) and pure H_2 (Fig. 2.21b) feed evidences how in the former case λ effect on cell voltage is less evident and, moreover, the switch from the first to the second effect zone corresponds to a different current density value.

2.3   Lumped (Gray-Box) Modeling

As deeply discussed in Chap. 1 and Sect. 2.1, the availability of fast and accurate dynamic models is nowadays considered highly strategic, especially to enhance the development of reliable and effective control and diagnostic strategies for SOFC systems. In this section, lumped modeling is proposed as a suitable approach to achieve a satisfactory compromise between computational time, experimental burden, and accuracy. The hierarchical sequence shown in Fig. 2.2 was fruitfully exploited to derive a control-oriented model of SOFC dynamic performance. Particularly, such a model was proven effective to study and the SOFC dynamic response to load change, thus deriving useful information towards the definition of real-world applicable management strategies for SOFC-based energy systems.

Furthermore, the lumped modeling approach was extended to the other main ancillary components, which an SOFC system consists of. Particularly, Fig. 2.22 highlights how achieving proper SOFC functioning entails surrounding the stack with several devices, the most significant ones being the air pre-heater, fuel pre-reformer, postburner, and air compressor. The following subsections present in detail the lumped submodels developed to enable dynamic simulation of an entire SOFC system. A larger attention was of course paid to the SOFC stack, for which not only the basic equations are presented and tested in simulation, but also

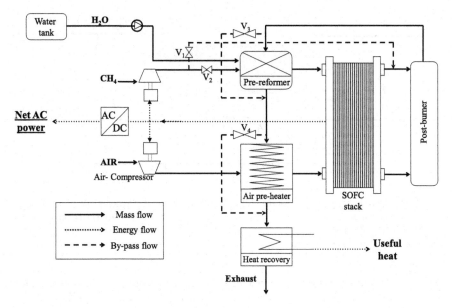

Fig. 2.22 Typical layout (adapted from Sorrentino and Pianese 2009) of an SOFC APU, showing the most significant ancillary devices along with principal energy and mass flows

experimental verification was carried-out to further confirm the high potential offered by lumped modeling approach for control and diagnostic purposes.

2.3.1 Lumped Model of Planar SOFC

The control-oriented model here described was conceived in such a way as to enable fast and accurate prediction of cell/stack dynamic response to any change in operating conditions. The model was developed assuming specific simplifying assumptions, listed below, whose validity is supported by previous studies:

- no heat exchange with the surroundings takes place, therefore adiabatic conditions are assumed (Braun 2002);
- gases sensible heat is neglected;
- pressure drop is assumed negligible (Burt et al. 2004);
- both electrochemistry and mass transfer are assumed as instantaneous processes, thus, thermal dynamics is the dominant phenomenon during transients (Achenbach 1995);
- the temperature of the solid trilayer is assumed as representative of the entire control volume (i.e., lumped modeling approach) (Iwata et al. 2000);
- a rectangular, planar, anode-supported, and co-flow SOFC is considered;

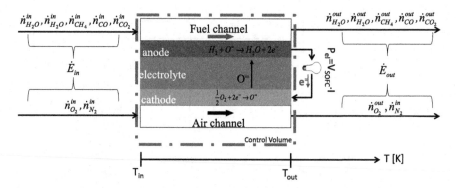

Fig. 2.23 Control volume (i.e., the one fully included within the *dotted-dashed* border) with reference to which the state-space model expressed by Eq. (2.58) was derived. Heat exchange with the surroundings was neglected as a consequence of adiabatic hypothesis, thus it was not included in the above illustration

- single SOFC dynamics is considered as representative of the entire stack behavior;
- T_{out} (see Fig. 2.23) is assumed as the state variable.

The above assumptions allow describing the dynamic behavior of an SOFC cell as a first-order system, with the outlet temperature being the only state variable. Particularly, the application of the first principle of Thermodynamics to the control volume shown in Fig. 2.23 results in the following state-space model:

$$
\begin{cases}
\dot{x} = \dfrac{dT_{out}}{dt} = f(x, u) = \dfrac{1}{K_{cell}} \left[\dot{E}_{in}(T_{in}) - \dot{E}_{out}(T_{out}) - \bar{J} \cdot V_{SOFC}(\bar{J}, T_{out}, U_f, \lambda) \cdot A \right] \\
y = V_{SOFC} = g(x, u)
\end{cases}
$$

$$(2.58)$$

where K_{cell} is the cell thermal capacity and $x = T_{out}$, $y = V_{SOFC}$ and $u = \{\bar{J}, T_{out}, U_f, \lambda\}$ are the state, output, and input variables, respectively. Particularly, K_{cell} is the key model parameter to be determined when specifically aiming at simulating transient SOFC response. The authors themselves previously discussed (Sorrentino et al. 2008) how the lumped capacity approach adopted to describe SOFC thermal dynamics (i.e., Eq. 2.58), would suggest to identify such a parameter against transient experiments. Nevertheless, geometrical-based estimation of K_{cell} also can be reliable.

The following subsection provides the reader with further details on how are estimated the energy flows leaving the control volume shown in Fig. 2.23 (i.e., $\dot{E}_{out}$ in Eq. 2.58).

It is also worth mentioning here that the state-space model output, i.e., V_{SOFC} in Eq. 2.58, is here computed by means of black-box submodels, which are developed according to the hierarchical sequence shown in Fig. 2.2. Later on in the chapter, specifically in Sects. 2.3.1.2 and 2.4.2, the two possible paths that can be potentially exploited to derive such a black-box relationship will be described and discussed in detail.

2.3.1.1 Estimation of Outlet Energy Flows

As shown in Fig. 2.23, the most general case here considered corresponds to a reformate fuel feeding resulting from methane pre-reforming. If this is the case, the outlet anode composition must be estimated to correctly evaluate the $\dot{E}_{out}$ term in Eq. (2.58). Here, a similar approach to the one proposed in Sect. 2.2.1.1 to estimate pre-reformer outlet composition was followed. Particularly, outlet molar flows can be estimated assuming that methane gets completely reformed inside the anode channels. Such a hypothesis well fits with experimental evidence and it was further demonstrated, in this work itself, through the analysis of estimated fuel composition along the gas channels (see Fig. 2.12).

The above hypothesis led to write the following system of four equations (C, H, and O balances plus the equilibrium constant evaluated for the water-gas-shift reaction) with four unknowns, namely all the terms having "out" as superscript:

$$\dot{n}_{CH_4}^{in} + \dot{n}_{CO}^{in} + \dot{n}_{CO_2}^{in} - \dot{n}_{CO}^{out} + \dot{n}_{CO_2}^{out} = 0 \tag{2.59a}$$

$$4\dot{n}_{CH_4}^{in} + 2\dot{n}_{H_2}^{in} + 2\dot{n}_{H_2O}^{in} - 2\dot{n}_{H_2}^{out} - 2\dot{n}_{H_2O}^{out} = 0 \tag{2.59b}$$

$$\dot{n}_{H_2O}^{in} + \dot{n}_{CO}^{in} + 2\dot{n}_{CO_2}^{in} + 2\dot{n}_{O_2,cons} - \dot{n}_{H_2O}^{out} - \dot{n}_{CO}^{out} - 2\dot{n}_{CO_2}^{out} = 0 \tag{2.59c}$$

$$K_{shift}(p, T_{out}) = \frac{\dot{n}_{H_2}^{out} \cdot \dot{n}_{CO_2}^{out}}{\dot{n}_{H_2O}^{out} \cdot \dot{n}_{CO}^{out}} \tag{2.59d}$$

where K_{shift} is evaluated by means of Eq. (2.4). As it emerges from the comparison between Eqs. (2.59a)–(2.59d) and (2.11a)–(2.11c) and (2.12), the major difference lies in the absence of the methane molar flow (i.e., $\dot{n}_{CH_4}^{out} = 0$) at the control volume outlet (shown on Fig. 2.23). A further difference corresponds to the fact that the oxygen molar flow, which crosses the electrolyte from cathode to anode to enable the hydrogen electro-oxidation (see Fig. 2.23), must be considered as a further inlet flow (i.e., $\dot{n}_{O_2,cons}$). Such oxygen consumption is accounted for when writing the oxygen balance, here expressed by Eq. 2.59c.

2.3.1.2 Analysis of Transient Response

The physical meaning of the lumped, control-oriented model presented above was tested by simulating the response of V_{SOFC} to step changes in load (i.e., average current density $\bar{J}$) and excess air λ. The interest for analyzing these particular responses is justified by the use of $\bar{J}$ and λ as control variables for matching the power requests and limiting temperature rise across the cell, respectively (Aguiar et al. 2004).

Below are defined the dynamic performance metrics used to study and discuss SOFC thermal response to step changes:

- $V_{drop} = V_{SOFC}(t_\infty) - \min[V_{SOFC}(t)]$: voltage undershoot intensity [mV];
- $\Delta V = V_{SOFC}(t_\infty) - V_{SOFC}(t_0)$: voltage difference between second (t_∞) and first (t_0) stationary point [mV];
- τ_V: voltage relaxation time [s];

Voltage relaxation time, which is a very significant variable to be accounted for when performing optimal design and balance of plant analysis for SOFC systems (Selimovic et al. 2005), is defined as the time needed for V_{SOFC} to recover up to 90 % of the voltage drop occurring after load or excess air step variation (Achenbach 1995):

$$V_{rec} = 0.9 \cdot V_{drop} + \min[V_{SOFC}(t)] \tag{2.60a}$$

$$\tau_V = t|_{V_{SOFC}=V_{rec}} - t_0 \tag{2.60b}$$

The cell geometry considered in this analysis is almost identical to the ES one specified in Table 2.2. In this specific analysis, anode, cathode and electrolyte thicknesses were assumed equal to 600, 50 and 10 µm, respectively, thus allowing to switch from an electrolyte- to a more up-to-date anode-supported cell. Upon knowledge of geometrical data and cell material (i.e., ceramic), the following geometrically derived value of cell capacity was assumed in the transient analysis of SOFC performance:

$$K_{cell} = \rho_{cer} \cdot VOL_{cell} \cdot c_{cer} = 6600 \left[\frac{kg}{m^3}\right] \cdot 2.08 \cdot 10^{-5} [m^3] \cdot 400 \left[\frac{J}{kg \cdot K}\right] = 55 \left[\frac{J}{K}\right]$$
$$\tag{2.61}$$

As for the estimation of V_{SOFC}, the following multiple linear regression (MLR) was used for a methane-fed, un-pressurized, planar, co-flow SOFC:

$$V_{SOFC} = g(x, u) = 0.18 - 0.08U_f - 1.25\bar{J} + 0.0041\lambda \frac{T_{out}}{1000}$$
$$+ 0.86J \frac{T_{out}}{1000} + 0.71T_{in} \tag{2.62}$$

Equation (2.62) was obtained adopting the hierarchical approach described in Sect. 2.1. Particularly, the blue-line path shown in Fig. 2.2 was applied as schematically described in Fig. 2.24: the 1D model presented in Sect. 2.2 was deployed to generate a set of virtual steady-state experiments, through which it was possible to well capture, as shown on the lower right corner of the figure, the

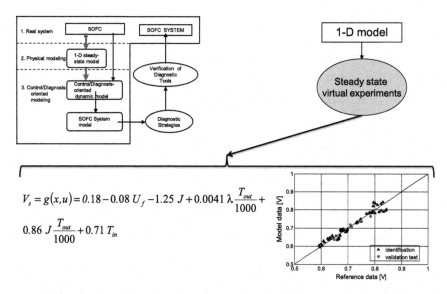

Fig. 2.24 Schematic description of hierarchical approach deployment for the development of the MLR correlation expressed by Eq. (2.62)

dependence of voltage on current density, inlet temperature, and fuel and air utilization (see Sect. 2.2.1.7).

Despite being a steady-state voltage submodel, the correlation expressed by Eq. (2.62) well adapts to the lumped, control-oriented model features. Indeed, voltage transient variation mainly depends on temperature, whose dynamics is assumed being the dominant one as in this work as in other relevant contributions devoted to SOFC dynamic behavior (Achenbach 1995; Haynes 2002). Therefore, the fact that Eq. (2.62) directly links voltage to inlet temperature and air utilization, which in turn affects inner and outer stack temperature, ensures satisfactory prediction capabilities both in steady-state and transient operations.

Response to Load Change

The SOFC response to load change was investigated for three different step variations (i.e,. case 1, case 2, and case 3, as shown in Fig. 2.25). The other input variables were set to the following nominal values:

- $T_{in} = 750\ ^{\circ}C$
- $\lambda = 6$
- $U_f = 0.8$

Figure 2.26 shows V_{SOFC} and T_{out} responses. For all the simulated cases, load increase causes polarization losses to sudden increase, resulting in an

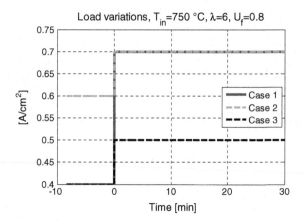

Fig. 2.25 Schematic illustration of load step changes considered in the response to load change analysis

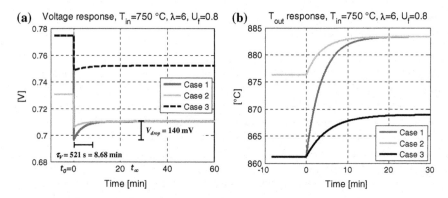

Fig. 2.26 Voltage (**a**) and outlet temperature (**b**) response to the load step-changes shown in Fig. 2.25. In (**a**) the values of t_0 an t_∞ resulting from cases 1 and 2 are indicated nearby the x-axis

undershoot-type voltage response, as shown in Fig. 2.26a. Such a behavior can be qualitatively explained considering the scheme shown in Fig. 2.27. At time $t = 0^-$, i.e., right before the load step change occurs, the voltage associated to current operating condition lies on the polarization curve associated to the given inlet temperature (i.e., $T_{in} = 750\ °C$). Then, at time $t = 0^+$, i.e., immediately after the step change in load, the operating condition not only suddenly moves toward the new current density, but also deviates from the previous I–V curve; this latter phenomenon is due to the fact that the line corresponding to $T_{in} = 750\ °C$ was obtained by modeling SOFC performance in steady-state conditions for current density varying between 0 and 1 A cm^{-2}. But when the load undergoes a step change, the subsequent thermal transient causes the temperature to slowly adapt to the next steady-state operating condition (see Fig. 2.27). Therefore, it can be assumed that at time $t = 0^+$ the new operating condition corresponds to a lower T_{in} level, which actually well emulates the low average stack temperature obtained right after the step change. From this point forward, thanks to the slow temperature increase shown in Fig. 2.26b, polarization losses decrease, thus allowing for a slow voltage recovery, which in turn causes the operating condition to slowly move back to the initial $T_{in} = 750$ °C I–V curve, following the vertical line linking the points labeled $t = 0^+$ and $t = t_\infty$, respectively (see Fig. 2.27).

Table 2.5 summarizes the results obtained by simulating cases 1, 2, and 3 (see Fig. 2.25). The values of τ_V indicate that thermal dynamics is significantly slow, of the order of hundreds of seconds, thus achieving satisfactory accordance with the simulations performed in previous works (Achenbach 1995; Lukas et al. 1999; Larrain 2005). Moreover, τ_V tends to reduce as $\bar{J}(t_\infty)$ increases. This behavior depends on the value of final current density reached after load step. In fact, the higher $\bar{J}(t_\infty)$, the more heat is released by the electrochemical reaction, resulting in a more significant temperature increase, as it emerges from Fig. 2.26b. Such an effect allows for faster compensation of polarization losses due to load increase,

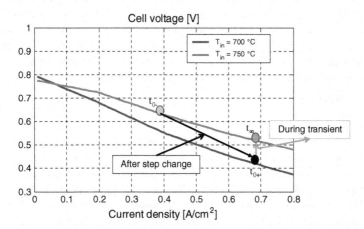

Fig. 2.27 Qualitative description of voltage evolution during a transient subsequent to a load step change

Table 2.5 Inputs and outputs of the response to load change analysis

Case	Step variables		Dynamic performance metrics	
	$\bar{J}(t_\infty) \left[\frac{A}{cm^2}\right]$	$\Delta\bar{J} = \bar{J}(t_\infty) - \bar{J}(t_0) \left[\frac{A}{cm^2}\right]$	τ_V (s)	V_{drop} (mV)
1	0.7	0.3	525.4	140
2	0.7	0.1	525.1	44
3	0.5	0.1	775.7	36

which in turn results in shorter τ_V for case 1 and case 2 as compared to case 3 (see Fig. 2.27 and Table 2.5).

The values of voltage drop, listed in Table 2.5, indicate a significant dependence of such a variable on both magnitude of load step (i.e., $\Delta \bar{J} = \bar{J}(t_\infty) - \bar{J}(t_0)$) and final current density (i.e. $\bar{J}(t_\infty)$). This can be explained considering that if $\bar{J}(t_\infty)$ is a few percent higher than $\bar{J}(t_0)$ (i.e. small $\Delta \bar{J}$ in cases 2 and 3, see Table 2.5 and Fig. 2.26a), steady V_{SOFC} and T_{out} variations are not so significant, thus resulting in small V_{drop}, as shown in Table 2.5. Moreover, the higher $\bar{J}(t_\infty)$, the larger difference between voltage values at different temperatures, as shown in Fig. 2.17. Thus, since immediately after the load step T_{out} remains still closer to $T_{\text{out}}(t_0)$ than $T_{\text{out}}(t_\infty)$, the higher $\bar{J}(t_\infty)$, the larger voltage drop.

Response to Excess of Air Change

An additional investigation was conducted on the voltage response to step changes in excess of air λ, which is commonly used as low-level control variable to keep the temperature increase from cell inlet to outlet in a safe range (i.e., $[100 \div 150]$ °C) (Zizelman et al. 2002). V_{SOFC} and T_{out} responses were simulated for the three cases listed in Table 2.6, each one referring to one of the three "effect-zones" (i.e., positive, no, and negative effect) highlighted in Fig. 2.21. Therefore, such an analysis allowed discussing the results with respect to the observations presented in the section devoted to steady-state parametric analyses. The other input variables were set to the following nominal values:

- $T_{\text{in}} = 750$ °C
- $U_f = 0.8$

Figures 2.29 and 2.30 show V_{SOFC} and T_{out} responses, respectively. Particularly, Fig. 2.28 indicate that, in all simulated cases, as λ increases T_{out} decreases, while V_{SOFC} always responds with an overshoot, which occurs because of the initial positive impact of higher amount of reactants (i.e., oxygen) in the cathode channel. Then, due to temperature reduction along the co-flow gas channels (see Fig. 2.29), V_{SOFC} slowly decreases down to the following stationary point (i.e., $V(t_\infty)$). As expected, current load $\bar{J}$, which is kept constant during the transient simulations

Table 2.6 Inputs and outputs of the response to change in excess air analysis

Case	Step variables			Dynamic performance metrics	
	$\bar{J} \left[\frac{A}{cm^2} \right]$	$\lambda(t_\infty)$	$\lambda(t_0)$	τ_V (s)	ΔV (mV)
4	0.1500	5	7	2466.2	4.7
5	0.3155			1100.9	0
6	0.7000			590.4	−13.4

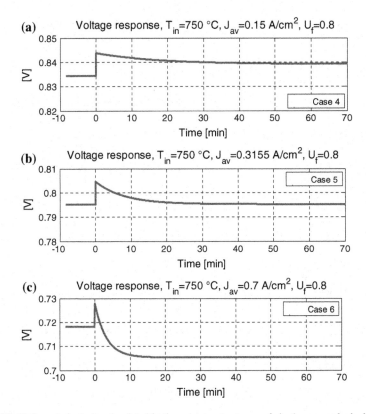

Fig. 2.28 Voltage trajectories simulated in the response to excess of air change analysis described in Table 2.6

here examined, strongly influences the response times, which decrease at high loads due to the higher amount of produced heat, as shown in Table 2.6.

Finally, as suggested by the steady-state parametric analysis carried-out in Sect. 2.2.1.7 (see Fig. 2.21), Table 2.6 and Fig. 2.28 show that ΔV varies from positive (i.e. case 4) to negative (i.e., case 6) values, as a function of the load. Therefore, this analysis not only allowed to investigate the dynamic effect of λ on cell performance, but also to further assess the accuracy of the black-box voltage correlation (i.e., Eq. 2.62) implemented in the lumped SOFC dynamic model.

2.3.1.3 Experimental Validation of the Lumped Model of Planar SOFC

The model presented and analyzed in the previous section was tested for experimental validation within the European project GENIUS (2013). Particularly, Eq. (2.58) was fed with all the inputs acquired by one of the industrial partners

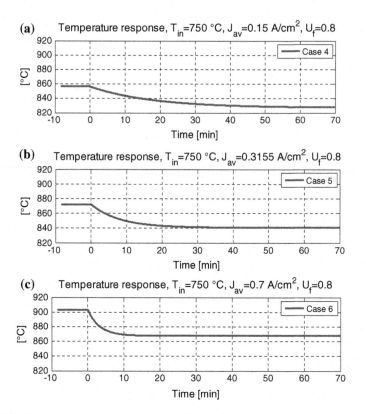

Fig. 2.29 Outlet temperature trajectories simulated in the response to excess of air change analysis described in Table 2.6

involved in the project, which include: inlet molar flows and temperature, voltage, and current density. Therefore, Eq. (2.58) allowed to estimate SOFC stack temperature and to compare it to the corresponding experimental measurement.

Figure 2.30 highlights the very good agreement achieved by extending the control-oriented modeling approach to a real system, thus confirming its suitability for on-field monitoring, control, and diagnostics of SOFC-based energy systems.

2.3.2 PostBurner Modeling

Figure 2.22 shows how the gases exhausting from anode and cathode channels are directly sent to the postburner, where they mix and react to produce additional heat for incoming fuel and air flows preheating and, eventually, to deliver useful heat when exiting the SOFC system. Moreover, since a highly toxic specie such as CO is

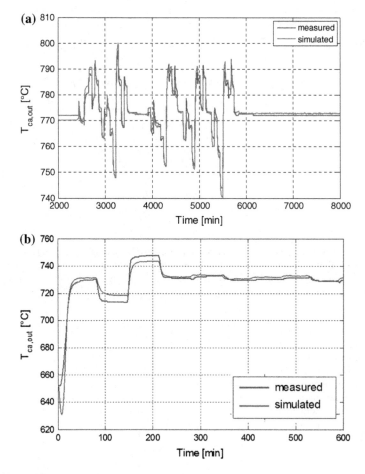

Fig. 2.30 Validation of the lumped model expressed by Eq. 2.58 against experimental transients consisting of a sequence of fast ramp changes in stack load

also present in the exhausts in a nonnegligible amount, its combustion contributes to lower the environmental impact of SOFCs.

Following the indications provided by Lu et al. (2006), the fact that cathodic channels are fed with extremely high excess of air (for SOFC internal cooling purposes) allows simplifying postburner modeling: it is therefore possible to assume that a complete combustion of H_2 and CO takes place inside this component, with a negligible impact of dissociation since outlet postburner temperatures are never so high (e.g., below 1400 °C). Moreover, the postburner is assumed adiabatic and its thermal dynamics considered negligible with respect to stack and heat-exchangers ones (Lu et al. 2006). Finally, the effect of combustion heat release rate can also be safely neglected. The above-introduced assumptions, as the ones that will be introduced later on for the heat exchangers (i.e., air preheater and steam

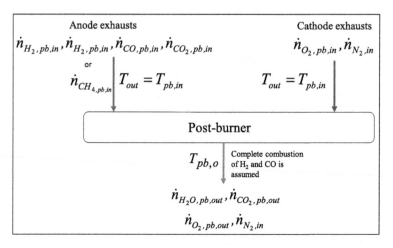

Fig. 2.31 Schematic representation of mass and energy flows entering and leaving the postburner used in a typical SOFC system (see Fig. 2.22). Note that nitrogen molar flows does never change going from SOFC system inlet to outlet, thus the corresponding notation did not change with respect to the control volume surrounding the SOFC stack, shown in Fig. 2.23

Fig. 2.32 Physical meaning of the energy balance over the postburner expressed by Eq. 2.63. The hypothesis of adiabatic flame temperature allows assuming that a constant enthalpy combustion process occurs inside the postburner (Heywood 1988)

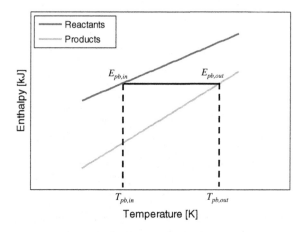

pre-reformer), are in accordance with the aim of developing a comprehensive control-oriented model for an SOFC system, to be placed at the lowest level of the hierarchical modeling structure shown in Fig. 2.2.

As a consequence of the above hypotheses, the process occurring within the postburner can be treated, with a reasonable approximation, by applying the adiabatic flame temperature estimation method (Heywood 1988) to the inlet and outlet flows shown in Fig. 2.31. Particularly, the outlet temperature $T_{pb,out}$ can be estimated by iteratively solving the following equation, which derives from the consideration that combustion products enthalpy must equal reactants enthalpy, as shown in Fig. 2.32:

$$\dot{E}_{\text{pb,in}}\left(T_{\text{pb,in}}\right) = \sum_{m} \dot{n}_{m,\text{pb,in}} \cdot h_m\left(T_{\text{pb,in}}\right) = \dot{E}_{\text{pb,out}}\left(T_{\text{pb,out}}\right)$$

$$= \sum_{m} \dot{n}_{m,\text{pb,out}} \cdot h_m\left(T_{\text{pb,out}}\right) m = [\text{H}_2, \text{H}_2\text{O}, \text{CH}_4, \text{CO}, \text{CO}_2, \text{O}_2, \text{N}_2]$$

$$(2.63)$$

where outlet molar flows are estimated as a function of inlet flows, through the following hydrogen (Eq. 2.64a), carbon (Eq. 2.64b) and oxygen balances (Eq. 2.64c). Equation 2.65 expresses the dependence of enthalpy on specific heat and enthalpy of formation of the generic species m:

$$\dot{n}_{\text{H}_2\text{O,pb,out}} = \dot{n}_{\text{H}_2\text{O,pb,in}} + \dot{n}_{\text{H}_2,\text{pb,in}} + 2 \cdot \dot{n}_{\text{CH}_4,\text{pb,in}} \qquad (2.64a)$$

$$\dot{n}_{\text{CO}_2,\text{pb,out}} = \dot{n}_{\text{CO}_2,\text{pb,in}} + \dot{n}_{\text{CO,pb,in}} + \dot{n}_{\text{CH}_4,\text{pb,in}} \qquad (2.64b)$$

$$\dot{n}_{\text{O}_2,\text{pb,out}} = \dot{n}_{\text{O}_2,\text{pb,in}} - \frac{1}{2}\dot{n}_{\text{H}_2,\text{pb,in}} - \frac{1}{2}\dot{n}_{\text{CO,pb,in}} - 2 \cdot \dot{n}_{\text{CH}_4,\text{pb,in}} \qquad (2.64c)$$

$$h_m(T) = \bar{h}^0_{f,m} + \int_{298}^{T} c_{p_m}(T)\mathrm{d}T \qquad (2.65)$$

It is finally worth here mentioning that the inlet flows shown in Fig. 2.31 are representative of both warmed-up and cold-start operations. Indeed, as it will be clarified in the next chapter, SOFC cold-start is here assumed being carried-out by directly feeding the postburner with methane; thus, in this specific working condition, the terms $\dot{n}_{\text{H}_2,\text{pb,in}}$, $\dot{n}_{\text{H}_2\text{O,pb,in}}$, $\dot{n}_{\text{CO,pb,in}}$ and $\dot{n}_{\text{CO}_2,\text{pb,in}}$ vanish in Fig. 2.31 and Eq. 2.64a–2.64c. On the other hand, during warmed-up phases, CH_4 normally gets completely reformed before leaving the SOFC anode channels (refer to the discussion made also in Sect. 2.3.1.1), thus leading to neglect the term $\dot{n}_{\text{CH}_4,\text{pb,in}}$ in such phases.

2.3.3   Heat Exchangers Modeling

Figure 2.22 indicates that there are two components, the air preheater and fuel pre-reformer, in which the hot gases exhausting from the postburner exchange heat with cold fluids, namely with the air and fuel flows entering the anodic and cathodic channels, respectively.

In order to cope with the purpose of developing a comprehensive dynamic model of SOFC systems, which is particularly required to fulfill the requirements, set at the control-oriented modeling level shown in Fig. 2.2, the zero capacity system approach proposed by Ataer et al. (1995) was here adopted. Such a modeling approach allows to describe, with a low computation effort, the response of heat exchangers to variation in hot, and cold fluid temperature and flows. Therefore, a good compromise, similar to the one attained for the SOFC stack (see Sect. 2.3.1), between accuracy and computational burden can be achieved. Moreover, a proper modeling balance is ensured when coupling different submodels with each other (e.g., SOFC stack and air preheater).

According to Ataer et al. (1995), the following simplifying hypotheses can be assumed:

- temperature changes linearly within both hot and cold channels;
- the heat exchanged between hot and cold fluid is proportional to the difference between the hot (i.e., $T_{h,\mathrm{HE}}$) and cold fluid (i.e., $T_{c,\mathrm{HE}}$) mean temperatures;

Therefore, the dynamics of both hot and cold fluid can be modeled as follows:

$$
\text{hot fluid:} \ (K_{\mathrm{HE}} + C_h) \frac{\mathrm{d}T_{h,\mathrm{HE}}}{\mathrm{d}t} = \dot{E}_{h,\mathrm{HE,in}}\left(T_{h,\mathrm{HE,in}}\right) - \dot{E}_{h,\mathrm{HE,out}}\left(T_{h,\mathrm{HE,out}}\right)
$$
$$
- U_{\mathrm{HE}} \cdot A_{\mathrm{HE}} \cdot \left(T_{h,\mathrm{HE}} - T_{c,\mathrm{HE}}\right)
$$
$$
\text{cold fluid:} \ C_c \frac{\mathrm{d}T_{c,\mathrm{HE}}}{\mathrm{d}t} = \dot{E}_{c,\mathrm{HE,in}}\left(T_{c,\mathrm{HE,in}}\right) - \dot{E}_{c,\mathrm{HE,out}}\left(T_{c,\mathrm{HE,out}}\right) \quad\quad (2.66)
$$
$$
+ U_{\mathrm{HE}} \cdot A_{\mathrm{HE}} \cdot \left(T_{h,\mathrm{HE}} - T_{c,\mathrm{HE}}\right)
$$

$$\mathrm{HE} = [\mathrm{aph}, \mathrm{pre}]$$

where the energy flow terms (i.e., $\dot{E}$) are estimated by applying Eq. (2.63) to the control volumes surrounding either the air presheater or the pre-reformer (see Fig. 2.22). Furthermore, the terms C_h and C_c represent the hot and cold fluid heat capacity, respectively.

The molar flows entering and leaving the pre-reformer are to be estimated as detailed in Sect. 2.3.1.1. It is worth remarking that the specific pre-reformer here considered, i.e., the one used in the configuration of Fig. 2.22, only consists of the evaporator and reactor units. This is possible because the heat, required both to transfer water from liquid to vapor phase and to enable the endothermic steam-reforming reaction, is supplied by the hot fluid (i.e., postburner exhausts), flowing through the pre-reformer device (Sorrentino and Pianese 2009; Jahn et al. 2005).

The following subsection focuses on the methodology adopted to design heat exchangers destined to SOFC system integration.

2.3.3.1 Heat Exchanger Design

Heat exchangers design consists in finding the appropriate transfer area to ensure that the targeted heat transfer from hot to cold fluid takes place. Following a widespread practice, the method proposed by Kays and London (1964) for compact heat exchangers was here adopted. Particularly, the key parameter to be evaluated is the heat exchanger effectiveness ε, defined as the ratio between effective heat transfer $\dot{Q}$ and the maximum heat transfer achievable by an ideal counterflow heat exchanger (Mastrullo et al. 1991):

$$\varepsilon = \frac{\dot{Q}}{\dot{Q}_{max}} = \frac{\dot{C}_h \cdot \Delta T_h}{\dot{C}_{min} \cdot \Delta T_{max}} = \frac{\dot{C}_c \cdot \Delta T_c}{\dot{C}_{min} \cdot \Delta T_{max}} \qquad (2.67a)$$

$$\Delta T_h = T_{h,\text{HE},in} - T_{h,\text{HE},out} \qquad (2.67b)$$

$$\Delta T_c = T_{c,\text{HE},out} - T_{c,\text{HE},in} \qquad (2.67c)$$

$$\Delta T_{max} = T_{h,\text{HE},in} - T_{c,\text{HE},in} \qquad (2.67d)$$

The terms $\dot{C}_c$ and $\dot{C}_h$ appearing at the right-hand side of Eq. (2.67a)–(2.67d) are estimated as the average heat capacity rate within the assigned temperature ranges. For the sake of brevity, Eq. 2.68 explains how such a calculation is performed for the sole hot fluid:

$$\dot{C}_h = \frac{\dot{Q}}{\left(T_{h,\text{HE},in} - T_{h,\text{HE},out}\right)}$$
$$= \frac{\sum_m \dot{n}_{m,h,\text{HE},in} \cdot h_m\left(T_{h,\text{HE},in}\right) - \sum_m \dot{n}_{m,h,\text{HE},out} \cdot h_m\left(T_{h,\text{HE},out}\right)}{\left(T_{h,\text{HE},in} - T_{h,\text{HE},out}\right)} \qquad (2.68)$$

Empirical correlations are available (Lee et al. 2011) to express the functional relationship linking effectiveness to key heat exchanger parameters, such as the average heat transfer coefficient U_{HE}, the transfer area A_{HE}, and the terms $\dot{C}_c$ and $\dot{C}_h$. Such relationships are illustrated in Fig. 2.33, where $\dot{C}_{min}$ and $\dot{C}_{max}$ are defined as the minimum and maximum value between $\dot{C}_c$ and $\dot{C}_h$, respectively. Particularly, Fig. 2.33 qualitatively describes the dependence $\varepsilon = f(A_{\text{HE}} \cdot U_{\text{HE}}/\dot{C}_{min}, \dot{C}_{min}/\dot{C}_{max})$ for a cross-flow heat exchanger, which is the type that most suitably fits the

Fig. 2.33 Variation of heat exchanger effectiveness as a function of both design and operating parameters (adapted from Mastrullo et al. 1991; Lee et al. 2011)

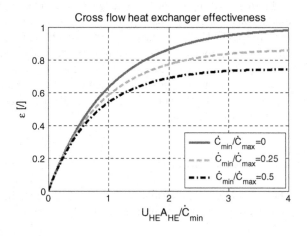

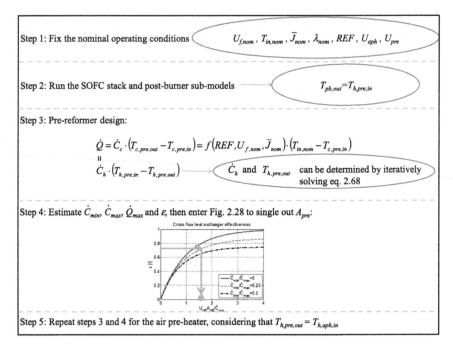

The diagram contains:

Step 1: Fix the nominal operating conditions $U_{f,nom}$, $T_{in,nom}$, $\bar{J}_{nom}$, λ_{nom}, REF, U_{aph}, U_{pre}

Step 2: Run the SOFC stack and post-burner sub-models → $T_{pb,out} = T_{h,pre,in}$

Step 3: Pre-reformer design:

$$\dot{Q} = \dot{C}_c \cdot \left(T_{c,pre,out} - T_{c,pre,in}\right) = f\left(REF, U_{f,nom}, \bar{J}_{nom}\right) \cdot \left(T_{in,nom} - T_{c,pre,in}\right)$$
$$\| $$
$$\dot{C}_h \cdot \left(T_{h,pre,in} - T_{h,pre,out}\right) \longrightarrow \dot{C}_h \text{ and } T_{h,pre,out} \text{ can be determined by iteratively solving eq. 2.68}$$

Step 4: Estimate $\dot{C}_{min}$, $\dot{C}_{max}$, $\dot{Q}_{max}$ and ε, then enter Fig. 2.28 to single out A_{pre}:

Step 5: Repeat steps 3 and 4 for the air pre-heater, considering that $T_{h,pre,out} = T_{h,aph,in}$

Fig. 2.34 Logic diagram synthesizing the model-based design of heat exchanger transfer area, for a given nominal operating condition established for the SOFC system

schematic layout of Fig. 2.22. Figure 2.33 also highlights that, once are assigned the heat transfer coefficient, the hot and cold fluid molar flows as well as the inlet and outlet temperatures, the values of effectiveness, and $\dot{C}_{min}/\dot{C}_{max}$ ratio can be computed through Eqs. (2.67a) and (2.68), thus allowing to enter Fig. 2.33 diagram and single out the current value of the abscissa $A_{HE} \cdot U_{HE}/\dot{C}_{min}$, which in turn can be used to estimate the transfer area.

Furthermore, upon the knowledge of the selected heat exchanger type (e.g., double pipe, shell and tube, plate, etc.), the surface to volume ratio can also be retrieved (NTNU 2013), thus allowing to determine heat exchanger volume. The latter variable, together with material properties knowledge, can be used to estimate (see Eq. 2.61) heat exchanger lumped capacity, whose evaluation is needed to correctly simulate temperature transients via Eq. (2.66).

The correct deployment of the method described above for an SOFC system is synthesized in Fig. 2.34.

2.3.3.2   Air Compressor

The air compression system has an important role in the design and control of SOFC systems. Particularly, it is required to supply air to the cathode in the right

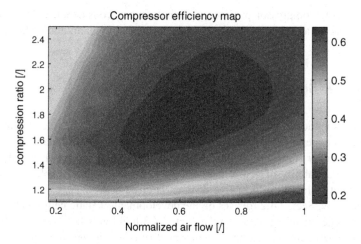

Fig. 2.35 Contour plot showing a normalized map linking compressor (i.e., Lyshom type, Miotti et al. 2006) efficiency to air flow and compression ratio

amount so as to guarantee, at the same time, the electrochemical reactions to take place appropriately and air-cooling of the SOFC stack to be safely performed.

Since the air compressor power represents the main loss within the entire SOFC system (see Fig. 2.22), its accurate estimation is needed to correctly estimate net performance. For the current study, a screw compressor (Miotti et al. 2006) was assumed. Compressor power is evaluated as a function of adiabatic efficiency and ideal compression work:

$$P_{cp} = \dot{m} \frac{c_p T_a}{\eta_{cp} \cdot \eta_{cm}} \left[\beta^{\frac{k-1}{k}} - 1 \right] \tag{2.69}$$

where β is the compression ratio, $\dot{m}$ is the inlet air flow processed by the compressor and η_{cp} and η_{cm} are the compressor and compressor motor efficiency, respectively. The last two efficiency terms are generally estimated via look-up tables, which are either supplied by manufacturers or developed by carrying-out suited experimental tests. Particularly, in this work η_{cp} is estimated as a function of mass flow and pressure ratio via an experimental efficiency map developed by (Miotti et al. 2006), as shown in Fig. 2.35.

Upon knowledge of compressor power adsorption via Eq. (2.69), net electric power yielded on output by the SOFC system (see Fig. 2.22) can be approximated as follows:

$$P_{net,AC} = P_{net,DC} \cdot \eta_{DC/AC} = \left(P_{gross} - P_{aux} \right) \cdot \eta_{DC/AC} \cong \left(P_{gross} - P_{cp} \right) \cdot \eta_{DC/AC}$$
$$= \left(n_{cells} \cdot V_{SOFC} \cdot I_{SOFC} - P_{cp} \right) \cdot \eta_{DC/AC}$$

$$\tag{2.70}$$

where the sole compressor parasitic contribution was considered, due to its much higher impact than other devices on net performance (Badami and Caldera 2002).

2.4 Black-Box Modeling

As already mentioned in Sect. 2.1.1 and schematized in Fig. 2.2, the availability of an extended experimental data set enables direct development of black-box models destined to either control or diagnostic algorithms development. Specifically in this section, the suitability of neural network (NN) models to fully exploit the knowledge content of highly informative data set is discussed.

Particularly, Sect. 2.4.2 focuses on the development of static neural networks for SOFC voltage modeling. Afterwards, Sect. 2.4.3 highlights how the availability of transient experiments enable the development of dynamic neural network models. The high accuracy granted by the latter models allows developing control algorithms that are as effective in transient maneuvers as in steady state, thus being more suitable than static black-box models for implementation in SOFC systems destined to load following applications (e.g., automotive, railway, marine propulsion, automotive, and airplane auxiliary power units).

2.4.1 Overview on Neural Network Models

Multi-layer-perceptron-feed-forward [MLPFF (Patterson 1995)] models are beyond doubt the most used artificial neural networks for fast and accurate steady-state prediction of highly nonlinear processes. Figure 2.36 provides a schematic representation of an MLPFF NN structure, highlighting the presence of nodes (i.e., processing units) and connections among nodes, to which the network parameters (i.e., the weights) are associated. For a deeper understanding of NN models features and basic theoretical principles, the reader is addressed to the abundant available literature (Patterson 1995; Haykin 1999; Ripley 2000). A brief overview on NN training and generalization is given below.

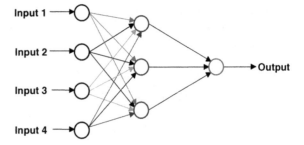

Fig. 2.36 Schematic representation of a generic multi-input single-output MLPFF NN structure. It is worth noting that the illustrated exemplary structure consists of four inputs, one hidden layer, and three hidden neurons

Network parameters are usually identified through a learning process, during which a set of training examples (generally experimental data) is fed to the NN to determine the weights of the connections linking network nodes. The most common approach adopted to identify NN parameters (i.e., connection weights) is the error backpropagation algorithm (Patterson 1995; Haykin 1999; Nelles 2000), due to its easy-to-handle implementation. At each iteration, the mean squared error (MSE, see Eq. 2.71) between experimental ($\hat{y}_l$) and NN output values (y_l) is propagated backward from the output to the input layer, passing through the hidden layers. The learning process is stopped when the MSE on a specified data set undergoes an assigned threshold.

$$\text{MSE} = \frac{1}{N} \sum_l (\hat{y}_l - y_l)^2 \qquad (2.71)$$

The training process mainly aims at determining NN models with a satisfactory compromise between accuracy and generalization. The latter feature can be maximized by selecting a training data set enough extended, so as to cover most of the system/process operating domain (Marra et al. 2013). Moreover, selecting an appropriate termination criterion for the backpropagation algorithm allows preventing network overfitting (Nelles 2000), which in turn may result in a significant loss of generalization. The early stopping is one of the most effective method against overfitting (Patterson 1995; Haykin 1999; Nelles 2000; Nørgaard et al. 2000). Particularly, the early stopping criterion consists in interrupting the training process once the MSE computed on a data set, different from the training one, stops decreasing. Therefore, when the early stopping is used, network training and test require at least three data sets (Haykin 1999): training-set (set A), early stopping test-set (set B) and generalization test-set (set C).

As for dynamic NN, the literature proposes a high number of effective neural structures that enable dynamic nonlinear modeling of many systems/processes. Since the analysis of all dynamic NN models falls out of the scope of the present book, a brief description of the adopted modeling methodology is here presented. Particularly, recurrent neural networks (RNN) can be obtained by suitably modifying the MLPFF NN structure shown in Fig. 2.36 (Nørgaard et al. 2000). The main modifications consist of: (i) feeding current and past input values, (ii) introducing feedback connections among the neurons, and thus providing the NN with a memory capability enabling dynamic simulations (see Fig. 2.37, where a typical RNN structure is shown).

Depending upon the feedback typology, which can either involves each neuron or only the output and input layers neurons, RNNs are classified into local RNN and global or external RNN, respectively (Haykin 1999; Nørgaard et al. 2000). Specifically in this work, the external RNN type was selected for dynamic modeling of SOFC performance, as discussed in Sect. 2.4.3. Detailed descriptions of external RNN structure and related implementation issues were given by the authors in

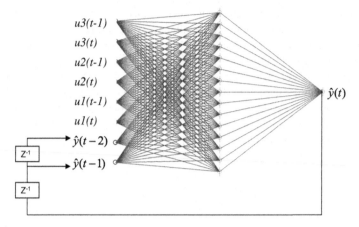

Fig. 2.37 Schematic representation of a generic RNN structure. It is worth noting that the illustrated exemplary structure consists of 3 inputs, 1 hidden layer and 15 hidden neurons. The memory effect, which provides dynamic simulation capabilities, is achieved by considering 1 past input and two output feedbacks $\hat{y}$

previous papers (Arsie et al. 2006, 2009, 2010), where the high potential of such models for control algorithm development in the automotive field was also demonstrated.

The neural network models developed and presented in the next two sections were trained applying the backpropagation algorithm in conjunction with the early stopping criterion, thus ensuring, as discussed above, both high accuracy and generalization. Moreover, trial and error analyses were carried-out to select NN structures with minimal number of parameters (weights), thus also allowing preventing network overparametrization, which can also result in generalization losses (Nelles 2000).

2.4.2 Development of Neural Networks for Steady-State Modeling of SOFC Performance

As shown in Fig. 2.2, the availability of a large number of experiments allows following the red path that directly links the real system, on which the experiments are gathered, to the control/diagnostics oriented modeling level. This case occurred in the framework of the GENIUS project (GENIUS 2013), which the authors participated in the timeframe 2010–2013. Particularly, several steady-state polarization curves were acquired on a fully experimented 5-cells short SOFC stack manufactured by Hexis, as described in a previous work (Marra et al. 2013).

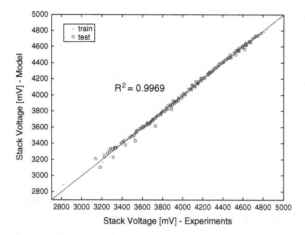

Fig. 2.38 Comparison between NN (i.e., model on the ordinates) outputs and measured voltage (i.e., experiments on the abscissa)

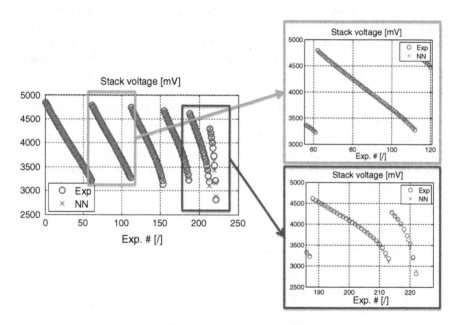

Fig. 2.39 Comparison of NN and experimental voltage over a voltage versus number of experiments domain

The stack was tested in such a way as to provide full I–V polarization curves at different operating conditions. For this specific application, an MLPFF NN (see Fig. 2.36) was selected to develop a black-box model linking voltage variation to main operating conditions, namely current density, inlet temperature, and anode and cathode inlet flows.

Figures 2.38 and 2.39 highlight the high accuracy guaranteed by such an NN model over the entire available data set. Particularly, the two close windows, shown at the right-hand side of Fig. 2.39, clearly highlight the very good precision ensured by neural network models in predicting steady-state SOFC performance as a function of load. For further details on NN structure definition, inputs selection and conceptual arguments associated to neural network training and testing, the reader is addressed to a previous authors' contribution (Marra et al. 2013).

2.4.3 Development of Neural Networks for Dynamic Modeling of SOFC Performance

As discussed above, the large number of experimental data, gathered within the GENIUS project (GENIUS 2013), enhanced the development of black-box models such as neural networks. Beyond the steady-state data acquired in the form of I–V profile (see Fig. 2.39), several transient voltage profiles were also acquired within the GENIUS project (GENIUS 2013), particularly on a stack manufactured by Topsoe Fuel Cell. Therefore, it was possible to develop RNN suitable to simulate SOFC performance both in steady-state and dynamic conditions, as discussed in Sect. 2.4.1.

Figure 2.40 shows the specific RNN structure developed to simulate SOFC voltage variation as a function of several input variables. Such an RNN configuration was deployed to simulate voltage profiles acquired on different SOFC systems, as detailed in a previous authors' contribution (Sorrentino et al. 2014), to which the reader is addressed for further details on network development and experimental activity.

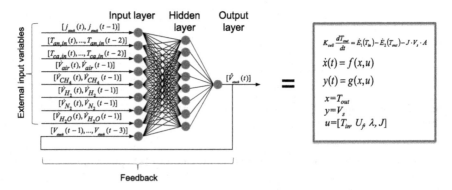

Fig. 2.40 RNN structure and qualitative equivalence between RNN and lumped models of SOFC dynamics

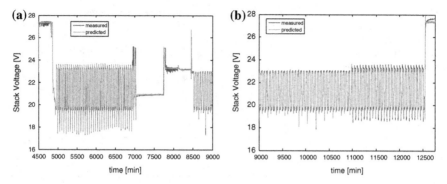

Fig. 2.41 Accuracy of the RNN SOFC performance simulator on the training (**a**) and test (**b**) data sets

The high accuracy and generalization granted by RNN for SOFC dynamic simulation is confirmed by the comparative analyses illustrated in Fig. 2.41. Particularly, the voltage dynamics induced by both step and ramp load variations can be well captured thanks to the modeling equivalence shown in Fig. 2.40. Indeed, the latter figure evidences how the capability of the RNN, of learning the thermal dynamics associated to main operating variables variation (refer to the discussions made in Sects. 2.3.1.2.1 and 2.3.1.2.2), allows to directly estimate dynamic effects, without the need of solving an ordinary differential equation, as it happens when applying a lumped modeling approach as the one proposed in Sect. 2.3.1. Such an aspect may positively lead to fast development of control-oriented models, which also meet the conflicting needs of high accuracy and low computational effort.

A drawback can be associated to the number of parameters to be stored on electronic control unit when identifying a complex neural model, whose parameters can amount up to 100–200 overall (Arsie et al. 2006). However, such an issue can be mitigated by applying network size reduction methods, such as the well-known optimal brain surgeon method (Nørgaard et al. 2000), which was proved, by the authors themselves (Arsie et al. 2013), very effective in reducing network size and, thus, over-parametrization problems.

2.5 Required Experimental Activity

Following the outcomes and observations generated by the numerical analyses presented above, the objective of this section is to provide useful recommendations on the required experimental activity, depending on both selected modeling approach and real-world application of developed models.

Fig. 2.42 Experimental
coverage (i.e. related to
black-points location) of
theoretically allowable
voltage operating domain

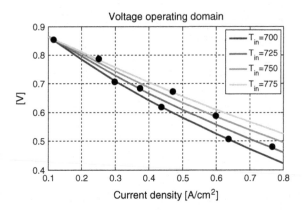

Fig. 2.42 Experimental coverage (i.e. related to black-points location) of theoretically allowable voltage operating domain

Particularly, Sect. 2.5.1 highlights the importance of fully covering the SOFC working domain, whenever the aim is to develop highly general steady-state models. On the other hand, Sect. 2.5.2 indicates how to select the most suitable transient maneuvers, when aiming at developing dynamical models of SOFC behavior.

2.5.1 Definition of Steady-State Identification Domain

Figure 2.42 shows a generic I–V domain, which qualitatively holds valid in representing the whole theoretically allowable working region of an SOFC. As it is well known, the relatively high fragility of SOFCs makes really hard to fully explore such a domain by experimental activities only. Nevertheless, accurate cell design, as well as materials selection, often entail evaluating cell performance and other variables (e.g., internal temperature, current, and composition distributions) in a considerable number of operating conditions.

One suitable solution to the above issues is to refer to the hierarchical modeling approach shown in Fig. 2.2. Indeed, the black-colored points shown in Fig. 2.42 could be used to map the external perimeter and most significant internal points of the SOFC working domain. Afterwards, a physical model (e.g., the one presented and discussed in Sect. 2.2.1) could be developed, thus allowing to perform design-oriented simulations in a larger domain.

Moreover, the developed physical model can be exploited to create virtually extended data sets. This corresponds nothing but to adopting the hierarchical approach associated to the blue line of Fig. 2.2, thanks to which the rest of the working domain of Fig. 2.42 can be covered by performing a large number of virtual experiments.

One possible application of the above-described virtual data set extension is the development of black-box steady-state models, such as the ones presented in Sect. 2.3.1 (see Fig. 2.24) and Sect. 2.4.2. Furthermore, the virtually extended data

Fig. 2.43 Schematic
representation of the transient
maneuvers described in
Table 2.7. Such a
representation is in
accordance with physical
response subsequent to a load
step change, as previously
discussed in Sect. 2.3.1.2.1

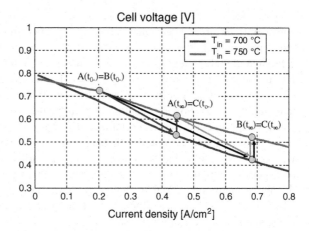

set can be used to identify inverse models of SOFC behavior, e.g., able to estimate
inlet fuel and air flows as a function of voltage and/or external temperature. Such
inverse models are potentially very useful to assess the effect of cell to cell tolerance, as a function of measured performance, or, eventually, to verify maldistri-
bution effects in multi-stack systems.

2.5.2   Transient Experiments

As deeply discussed in Sect. 2.3.1.2, transient response of SOFC voltage to load
change strongly depends on the conditions before and after the transient maneuver.
Figure 2.43 shows three possible load step-changes on a generic I–V domain.
Particularly, maneuver A starts with low current density and ends with medium
current, whereas maneuvers B and C end with high current density. The difference
between maneuvers B and C lies in the fact that the former starts with a low current,
while the latter with a medium current.

Table 2.7 Inputs and outputs of the response to load changes illustrated in Fig. 2.43

Case	Step variables		Dynamic performance metrics	
	$\bar{J}(t_\infty)\left[\frac{A}{cm^2}\right]$	$\Delta\bar{J} = \bar{J}(t_\infty) - \bar{J}(t_0)\left[\frac{A}{cm^2}\right]$	τ_V (s)	V_{drop} (mV)
A	0.45	0.25	866.6	8.7
B	0.7	0.5	521.4	24.7
C	0.7	0.25	521.1	11.5

The simulations were carried-out assuming the same cell geometry and operating conditions of
Sect. 2.3.1

The effect of different maneuvers on voltage response, which is based on the considerations already deepened in Sect. 2.3.1.2, is resumed by Table 2.7. The latter table must be considered of particular relevance when designing an experimental campaign aimed at identifying dynamic models of SOFC performance. Indeed, the fact that equal final current density values (i.e., cases B and C) are associated to equal voltage relaxation times and significantly different voltage drops (i.e., B more than doubles C, as shown in Table 2.7), clearly addresses the need for running quite various experimental transients. Particularly, not only most of the current density range must be explored, but also transient maneuvers have to be imposed in such a way as to ensure different load change as well as various initial and final load values be experimentally investigated. Of course the higher the number of combinations of the above mentioned variables values, the more informative the transient data set, which in turn will increase the generalization capabilities of black-box dynamic models, such as the RNN presented in Sect. 2.4.3.

2.6 Chapter Closure

This chapter presented in detail a hierarchical modeling structure, which is suitable to tackle main practical design problems associated to SOFC development and commercial success, namely system sizing as well as control and diagnostic strategies definition.

The compromise between physical content and computational burden of phenomenological and gray- and black-models was addressed and discussed with respect to their real-world potential deployment. Particularly, it was explained how the use of dimensional models is propaedeutic to performing parametric and scenario analyses of practical relevance, as well as to provide control-oriented model developers with key physical information, to be accounted for when selecting and developing less physical models. The latter modeling approaches were in turn proven effective as mathematical tools to be referred to when aiming at performing optimal balance of plant analyses, as well as single component sizing as a function of desired performance and fuel cell system nominal power.

All SOFC stack models were tested for validation against both reference numerical benchmarks and experiments. Numerical benchmarks were particularly useful in demonstrating the general validity and significant reliability granted by the developed 1D model. Particularly, main SOFC performance (e.g., voltage and fuel consumption), as well as spatial variations of most relevant operating variables, such as current density, temperatures, and species compositions, were accurately estimated. On the other hand, experiments gathered on different stacks confirmed the high potential offered by both black-box (including neural networks) and gray-box (e.g., lumped capacity models) to simulate voltage and outlet temperature in both steady-state and transient conditions.

The following two chapters focus on main methodologies and practical suggestions, to be accounted for when applying the proposed models hierarchy for model-based development of effective and reliable control and diagnostic tools.

References

Achenbach E (1995) Response of a solid oxide fuel cell to load change. J Power Sources 57:105–109

Achenbach E, Riensche E (1994) Methane/steam reforming kinetics for solid oxide fuel cells. J Power Sources 52:283–288

Aguiar P, Adjiman CS, Brandon NP (2004) Anode-supported intermediate temperature direct internal reforming solid oxide fuel cell. I: model-based steady-state performance. J Power Sources 138:120–136

Aguiar P, Adjiman CS, Brandon NP (2005) Anode-supported intermediate-temperature direct internal reforming solid oxide fuel cell: II. Model-based dynamic performance and control. J Power Sources 147:136–147

Andersson D, Åberg E, Eborn J (2011) Dynamic modeling of a solid oxide fuel cell system in Modelica. In: Proceedings of 8th Modelica conference, Dresden, Germany, 20–22 Mar 2011

Arsie I, Pianese C, Rizzo G, Flora R, Serra G (1999) A hierarchical system of models for the optimal design of control strategies in spark ignition automotive engines. In: Proceedings of 14th IFAC world congress, Beijing (China), July 5–9, pp 473–488

Arsie I, Pianese C, Sorrentino M (2006) A procedure to enhance identification of recurrent neural networks for simulating air-fuel ratio dynamics in SI engines. Eng Appl Artif Intell 19:65–77

Arsie I, Pianese C, Sorrentino M (2009) Development and real-time implementation of recurrent neural networks for AFR prediction and control. SAE Int J Passeng Cars Electron Electr Syst 1:403–412

Arsie I, Di Domenico A, Pianese C, Sorrentino M (2010a) A multilevel approach to the energy management of an automotive polymer electrolyte membrane fuel cell system. ASME Trans J Fuel Cell Sci Technol 7:0110041–01100411

Arsie I, Pianese C, Sorrentino M (2010b) Development of recurrent neural networks for virtual sensing of NOx emissions in internal combustion engines. SAE Int J Fuels Lubricants 2:354–361

Arsie I, Cricchio A, De Cesare M, Pianese C, Sorrentino M (2013) A methodology to enhance design and on-board application of neural network models for virtual sensing of NOx emissions in automotive diesel engines. In: Proceedings of the 11th international conference on engines & vehicles, Capri (Napoli), Italy, 15–19 Sept 2013. doi:10.4271/2013-24-0138

Ataer OE, Ileri A, Gogus Y (1995) Transient behavior of finned-tube cross-flow heat exchangers. Int J Refrig 18:153–160

Badami M, Caldera C (2002) Dynamic model of a load-following fuel cell vehicle: impact of the air system. SAE Technical Paper 2002-01-0100. doi:10.4271/2002-01-0100

Blum L, de Haart LGJ, Malzbender J, Menzler NH, Remmel J, Steinberger-Wilckens R (2013) Recent results in Jülich solid oxide fuel cell technology development. J Power Sources 241:477–485

Braun RJ (2002) Optimal design and operation of solid oxide fuel cell systems for small-scale stationary applications. Ph.D. thesis, University of Wisconsin, Madison, WI

Burt AC, Celik IB, Gemmen RS, Smirnov AV (2004) A numerical study of cell-to-cell variations in a SOFC stack. J Power Sources 126:76–87

Chan SH, Khor KA, Xia ZT (2001) A complete polarization model of a solid oxide fuel cell and its sensitivity to the change of cell component thickness. J Power Sources 93:130–140

Chick LA, Williford RE, Stevenson JW (2003) Spreadsheet model of SOFC electrochemical performance. SECA modeling & simulation training session. Available at http://www.netl.doe. gov/publications/proceedings/03/seca-model/Chick8-29-03.pdf

Ferguson JR, Fiard JM, Herbin R (1996) Three-dimensional numerical simulation for various geometries of solid oxide fuel cells. J Power Sources 58:109–122

GENIUS (2013) GEneric diagNosis InstrUment for SOFC systems. https://genius.eifer.uni-karlsruhe.de/

Guezennec Y, Choi TY, Paganelli G, Rizzoni G (2003) Supervisory control of fuel cell vehicles and its link to overall system efficiency and low-level control requirements. In: Proceedings of the 2003 American control conference, vol 3. June 4–6, Denver, CO (USA), pp 2055–2061

Haykin S (1999) Neural networks. Prentice-Hall, Englewood Cliffs

Haynes C (1999) Simulation of tubular solid oxide fuel cell behavior for integration into gas turbine cycles. Ph.D. thesis, Georgia Institute of Technology, Atlanta, GA

Haynes C (2002) Simulating process settings for unslaved SOFC response to increases in load demand. J Power Sources 109:365–376

Heywood JB (1988) Internal combustion engine fundamentals. MC Graw Hill, USA, pp 80–94

Iwata M, Hikosaka T, Morita M, Iwanari T, Ito K, Onda K, Esaki Y, Sakaki Y, Nagata S (2000) Performance analysis of planar-type unit SOFC considering current and temperature distributions. Solid State Ionics 132:297–308

Jahn HJ, Schroer W (2005) Dynamic simulation model of a steam reformer for a residential fuel cell power plant. J Power Sources 150:101–109

Kays WM, London AL (1964) Compact heat exchanger, 2nd edn. McGraw-Hill, New York

Larminie J, Dicks A (2003) Fuel cell systems explained. Wiley, Chichester, pp 1–24, 207–227

Larrain D (2005) Solid oxide fuel cell stack simulation and optimization, including experimental validation and transient behavior. Ph.D. thesis, École Polytechnique Fédérale de Lausanne (France)

Larrain D, Van HJ, Maréchal F, Favrat D (2003) Thermal modeling of a small anode supported solid oxide fuel cell. J Power Sources 138:367–374

Lee TS, Chung JN, Chen YC (2011) Design and optimization of a combined fuel reforming and solid oxide fuel cell system with anode off-gas recycling. Energy Convers Manag 52:3214–3226

Lu N, Li Q, Sun X, Khaleel MA (2006) The modeling of a standalone solid-oxide fuel cell auxiliary power unit. J Power Sources 161:938–948

Lukas MD, Lee KY, Ghezel-Ayagh H (1999) Development of a stack simulation model for control study on direct reforming molten carbonate fuel cell power plant. IEEE Trans Energy Convers 14:1651–1657

Marra D, Sorrentino M. Pianese C, Iwanschitz B (2013) A neural network estimator of Solid Oxide Fuel Cell performance for on-field diagnostics and prognostics applications. J Power Sources. 241:320–329

Massardo AF, Lubelli F (2000) Internal reforming solid oxide fuel cell—gas turbine combined cycles (IRSOFC-GT). Part A: cell model and cycle thermodynamic analysis. ASME Trans J Eng Gas Turbines Power 122:27–35

Mastrullo R, Mazzei P, Naso V, Vanoli R (1991) Fondamenti di trasmissione del calore (in Italian), vol 1. Liguori Editore, Napoli, Italy, pp 263–271

Miotti A, Di Domenico A, Esposito A, Guezennec YG (2006) Transient analysis and modelling of automotive PEM fuel cell system accounting for water transport dynamics. In: Proceedings of the fourth ASME international conference on fuel cell science, engineering and technology, Irvine, California, USA, 19–21 June 2006

Nelles O (2000) Nonlinear System identification. Springer, Berlin

Noren DA, Hoffman MA (2005) Clarifying the Butler-Volmer equation and related approximations for calculating activation losses in solid oxide fuel cell models. J Power Sources 152:175–181

Nørgaard M, Ravn O, Poulsen NL, Hansen LK (2000) Neural networks for modelling and control of dynamic systems. Springer, London

NTNU (Norwegian University of Science and Technology) (2013) Guide to Compact heat exchangers. Available at http://www.ntnu.no/ept/

Ormerod RM (2003) Solid oxide fuel cells. Chem Soc Rev 32:17–28

Patterson DW (1995) Artificial neural networks—theory and applications. Prentice-Hall, Englewood Cliffs

Pukrushpan JT, Stefanopoulou AG, Peng H (2004) Control of fuel cell breathing. IEEE Control Syst Mag 24(2):30–46

Ramallo-González AP, Eames ME, Coley DA (2013) Lumped parameter models for building thermal modelling: an analytic approach to simplifying complex multi-layered constructions. Energy Build 60:174–184

Ripley BD (2000) Pattern recognition and neural networks. Cambridge University Press, Cambridge

Rizzoni G, Josephson JR, Soliman A, Hubert C, Cantemir CG, Dembski N, Pisu P, Mikesell D, Serrao L, Russell J, Carroll M (2005) Modeling, simulation, and concept design for hybrid-electric medium-size military trucks. In: Trevisani DA, Sisti AF (eds) Proceedings of SPIE. Enabling technologies for simulation science IX, vol 5805, pp 1–12

Romijn R, Özkan L, Weiland S, Ludlage J, Marquardt W (2008) A grey-box modeling approach for the reduction of nonlinear systems. J Process Control 18:906–914

Selimovic A, Kemm M, Torisson T, Assadi M (2005) Steady state and transient thermal stress analysis in planar solid oxide fuel cells. J Power Sources 145:463–469

Sofcpower (2013) Sofcpower cells. Available at http://www.sofcpower.com/13/cells.html

Sorrentino M, Pianese C (2009) Control oriented modeling of solid oxide fuel cell auxiliary power unit for transportation applications. ASME Trans J Fuel Cell Sci Technol 6:041011–04101112

Sorrentino M, Mandourah AY, Petersen TF, Guezennec YG, Moran MJ, Rizzoni G (2004) A 1-D planar solid oxide fuel cell model for simulation of SOFC-based energy systems. In: Proceedings of 2004 ASME IMECE, Anaheim, California, USA, pp 205–214, 13–19 Nov 2004

Sorrentino M, Pianese C, Guezennec YG (2008) A hierarchical modeling approach to the simulation and control of planar solid oxide fuel cells. J Power Sources 180:380–392

Sorrentino M, Marra D, Pianese C, Guida M, Postiglione F, Wang K, Pohjoranta A (2014) On the use of neural networks and statistical tools for nonlinear modeling and on-field diagnosis of solid oxide fuel cell stacks. Energy Procedia 45:298–307

Topsoe Fuel Cell (2013) Topsoe fuel cell brochure. Available at http://www.topsoefuelcell.com/news_and_info/press_kit.aspx

U.S. Department of Energy (2004) Fuel cell handbook—seventh edition. Available at http://www.netl.doe.gov/technologies/coalpower/fuelcells/seca/refshelf.html

Yang X (2013) A higher-order Levenberg–Marquardt method for nonlinear equations. Appl Math Comput 219:10682–10694

Yang S, Chen T, Wang Y, Peng Z, Wang WG (2013) Electrochemical analysis of an anode-supported SOFC. Int J Electrochem Sci 8:2330–2344

Zizelman J, Shaffer S, Mukerjee S (2002) Solid oxide fuel cell auxiliary power unit—a development update. SAE Paper 2002-01-0411

Chapter 3
Models for Control Applications

3.1 Multilevel Control of SOFC Systems

Complex systems such as SOFC units, in which several energy and heat exchanges take place and, furthermore, mutual interactions linked to mass flows occur among different components, entail developing suitable control architectures. Such architectures are particularly required to optimize global energy efficiency, while guaranteeing proper and safe operation of main devices (e.g., the fuel cell stack), as well as of all system ancillaries (e.g., blowers/compressors, heat exchangers, and so on).

Designing a controller for an SOFC unit has to address the following two main issues: (i) to efficiently follow the target load, and (ii) to meet the constraints set by the different components safe operation limit. While the primary objective is to deliver the desired power at optimal efficiency, it is important to operate the fuel cell within certain limits. Since the operating temperature of SOFCs is higher than other fuel cells, the thermal stress among different components due to different temperature distributions along the anode, cathode, and electrolyte can be fatal to the integrity of the entire fuel cell. Thus it is of great importance to keep the maximum cell temperature below a certain threshold, as well as to reduce the temperature gradient. One of the main bottlenecks limiting SOFC lifetime is temperature gradient and hot spots of the cells that can cause cell damage. The analysis of temperature distribution inside the cell needs at least one-dimensional model (as discussed in Chaps. 1 and 2). An optimal temperature gradient control would be to maintain temperature profile of the cell as much uniform as possible. Apart from the physical constraints of the SOFC itself, limits on fuel and air utilization, maximum rate of load change, and technical constraints of the balance of plant (BoP) components play a key role in the operation of the SOFC and thus in the control design. For example, the lower limit imposed on the steam to carbon ratio at the entrance of the pre-reformer to avoid carbon deposition can be a limiting factor in the operation of the SOFC (Aguiar et al. 2005). Carbon deposition can change anode chemical composition as well as electrochemical reaction rate, and thus

© Springer-Verlag London 2016
D. Marra et al., *Models for Solid Oxide Fuel Cell Systems*,
Green Energy and Technology, DOI 10.1007/978-1-4471-5658-1_3

significantly reduce SOFC performance and lifetime. Therefore, preventing SOFC from carbon deposition is one of the important factors to be considered in control design (Weber et al. 2002).

These objectives can be fulfilled by developing different control strategies, which can be classified according to the three control levels shown in Fig. 3.1. The level-based organization of control rules, which of course is linked to control targets ranging from supervisory-level (e.g., global system efficiency) to low-level goals (e.g., operating the stack at the desired condition), was already addressed as a very effective approach for fuel cell hybrid units (Guezennec et al. 2003; Arsie et al. 2007).

A multilevel control framework usually includes supervisory-, central-, and low-level control strategies, as shown in Fig. 3.1. Supervisory control aims at appropriately managing available energy conversion devices (e.g., fuel cell system, including stack and ancillaries, and batteries) as a function of their operating features (i.e., efficiency curve) and health status (e.g., operating hours, depth of discharge (DOD) cycles, and so on). The final objective is to minimize the consumption of the energy primary source (e.g., methane) and meet load demand. Therefore, the supervisory-level is not directly involved in the fulfillment of control tasks, but mostly focuses on optimizing global efficiency by assigning reference guidelines to lower levels. Obviously, supervisory strategies depend on both specific system features and control objectives. Particularly, either in a hybrid (e.g., with batteries) stand-alone fuel cell system (FCS) or in grid-connected ones, the supervisory-level serves at properly sharing the load among fuel cell system and hybridizing devices and/or the grid. Moreover, a charge sustaining strategy must be guaranteed for the battery pack (if present), within either the load demand time window or the entire day. In the latter case, battery recharging through the grid must be introduced. As a consequence, a supervisory energy management policy has to be defined so as to enable the following operating modes: (i) only the main energy

Fig. 3.1 General description of multilevel approach to control strategies definition and implementation in SOFC energy units

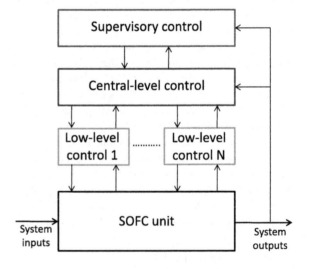

conversion device (e.g., the fuel cell system) meets load demand; (ii) both fuel cell system and batteries (and or grid) supply electric energy, thus resulting in a hybrid mode operation; (iii) the main energy device supplies energy for load and batteries recharging; and (iv) only batteries meet load demand. The above modes identify the energy flows combination within the entire hybrid unit (e.g., SOFC system plus batteries), as sketched in Fig. 3.2.

Central-level control serves at making low-level control actions suitably match supervisory guidelines, as introduced before. The central-level defines the working set points for each system component and properly regulates the transition from one mode to the other. In this way, such a control level will enable efficient operation of the entire energy conversion system, while ensuring safe and healthy functioning of all components. Final outcome of such a control level is to provide the set points to be tracked by system actuators at the inferior level.

The objective of the low-level control is to make the FCS work at the desired set points, as addressed by the central-level, while guaranteeing proper transient response. The low-level control drives the physical actuators of the system under investigation, accounting for the main dynamics involved in the energy, heat exchange, and mass transfer processes. Differently than the above levels, which

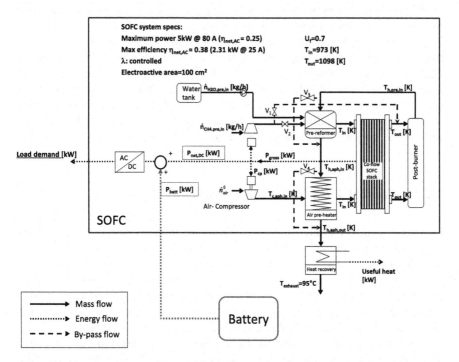

Fig. 3.2 Schematic layout of a hybrid SOFC unit, which also includes main system specifications. Further details on the specific model-based design methodology, used to size the SOFC system, are retrievable from Sorrentino and Pianese (2009). The variables shown within the SOFC system box refer to the components sub-models introduced in Chap. 2

only address what are the reference set points to be traced during system operation, at the low-level, all the effective and on-field deployable control rules and physical controllers (e.g., Proportional–integral–derivative, PID) are to be designed and suitably developed. Of course, the overall multilevel architecture must be designed in such a way as to guarantee optimal interfacing, through the central-level, between supervisory- and low-level controllers. Indeed, from a general point of view it must be always recalled how increasing the complexity of adopted control strategies always causes the three control levels to superpose each another, thus resulting in a unique, vertically integrated control architecture.

Specifically, in this chapter, a multilevel approach is proposed to successfully perform model-based development and real-world implementation of optimal energy management strategies for the SOFC hybrid unit shown in Fig. 3.2. Particularly, the following Sect. 3.1.1 provides a synthetic overview on main control variables and controlled outputs, to be accounted for when developing a control architecture for SOFC systems. Then, Sect. 3.1.2 briefly reports on the most suitable controllers to be adopted for facing the energy and thermal management issues described in Chap. 1. Sections 3.2 through 3.4 provide details on model-based definition of control rules at each level (see Fig. 3.1). Finally, Sect. 3.5 presents and discusses a case study, focusing on the deployment of the hybrid SOFC unit here considered both as an automotive auxiliary power unit (APU) and for stationary combined heat and power generation.

It is worth here remarking how the SOFC unit shown in Fig. 3.2, whose model-based design was previously performed in Sorrentino and Pianese (2009), was selected as the reference system for the following descriptions, due to its simplicity and general-purpose features. Moreover, the rules defined at the supervisory-level (see Sect. 3.2) will be shown to be effective for batteries as well as for other relevant hybridizing devices and, eventually, for sharing the load demand with the electricity supplier in grid-connected SOFC systems.

3.1.1 Control Variables and Controlled Outputs

Figure 3.3 synthetically illustrates the main control variables and controlled outputs, to be accounted for when developing an appropriate control structure for the selected SOFC system. It is worth noting that the variables shown in Fig. 3.3 only refer to the low-level control part of the general overall architecture presented in the previous section. Indeed, it is important to clarify that high-level controllers process external requests in such a way as to convert them into suitable set points, to be then referred to by low-level controllers, as described in the above section.

Specifically, the load demand has to be treated as an external disturbance, since the objective is of course to match the electricity request. Nevertheless, such a load demand is suitably processed by the multilevel control architecture (see Sect. 3.2 through Sect. 3.4). The latter processing results in fixing the current at which the SOFC stack has to be operated to match the power load demand (see Fig. 3.2), thus

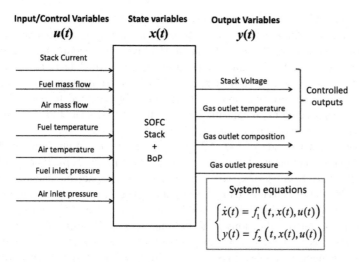

Fig. 3.3 Schematic summary of main control variables and controlled outputs involved in SOFC units functioning

explaining why *I* is considered as a low-level control variable hereinafter. In any case, the reader is referred to the subsequent sections to get a further and clearer insight on such an aspect.

Figure 3.3 groups at the left-hand side of the input variables, which also correspond to the main control variables to be managed by FCS actuators. Particularly, a suitable DC/DC power conditioning device is required not only to perform good coupling between battery and SOFC stack, but also to make the SOFC system work at the desired *I* value (Valdivia et al. 2014), as addressed by the low-level current controller (see Figs. 3.10 and 3.11). Air and fuel flows are temperature-related control variables, due to their direct impact on stack and outlet temperatures. In this case, the control action is obtained by suitably regulating the air and fuel compressors/fans shown in Fig. 3.2. Regarding the pressure level of incoming flows, it is accounted for among input/control variables to well represent all possible operations of SOFC stacks, which indeed can be also operated in pressurized conditions. In the latter case, the inlet air and fuel pressures will considerably impact on voltage and performance, whereas in the most common case of unpressurized operation, their regulation must be managed aiming at compensating for pressure losses occurring inside the fuel cell channels.

On the right-hand side of Fig. 3.2, the main controlled outputs correspond to SOFC voltage and outlet temperature. The interests for performing accurate control of such variables are linked to their effect on SOFC performance and efficiency, as deeply discussed in the previous chapter. It is worth recalling here how, theoretically, best performance can be achieved by increasing as much as possible operating temperature. Nevertheless, practical considerations linked to SOFC safety and lifetime suggest to limit temperature increase across cell length (Aguiar et al. 2005). In this context, the availability of a control-oriented model, as the one described in

Sect. 2.3, is really desired, as it enables performing advanced model-based design analysis aimed at finding the best compromise among system efficiency and safe operation. Particularly, the model can be deployed to account for the benefits achievable by letting SOFC outlet temperature increase, which has a twofold impact on efficiency: (i) gross efficiency increases with temperature, as shown in Figs. 2.17 and 2.18 and (ii) a reduced amount of air has to be supplied to the cathode for cooling, thus reducing main parasitic loss, i.e., the one associated to air compressor power adsorption, as discussed in Sect. 2.3.3.

On the right-hand side of Fig. 3.3 further outputs are shown. Particularly, gas composition is interesting to monitor, mainly due to its impact on postburner functioning (see Sect. 2.3.2), which in turn has an impact on balance of plant analyses. The latter activities can be also enhanced by deploying the model-based approach to optimize both inlet and outlet fuel compositions as well as reformer characteristics and specifications, with the final aim of determining the best components sizing as a function of end applications. For instance, when applying SOFC units for cogeneration applications, the heat produced by the postburner, which of course will directly impact on final thermal energy delivered on output, must be carefully accounted for.

3.1.2 Which Controllers?

The aim of this section is to briefly recall the most significant control approaches that can be deployed to attain optimal and safe operation of SOFC units. For a detailed analysis of the state of the art and rigorous understanding of how to address control problems, depending both on application field and performance and cost targets, the reader is addressed to the abundant literature available in the public domain (Franklin et al. 2006; Bolton 2002).

Figure 3.4a highlights two main control configurations, namely open-loop (i.e., feedforward control) and closed-loop (i.e., feedback control). In the former case, the control action, to be performed by system actuators as a function of the control signal yielded on output by the selected controller (for instance by a look-up table or a map, as shown in Fig. 3.4b), is exerted independently from its effect on system outputs. On the other hand, feedback controllers impose the control action as a function of the error between desired target and current value of the controlled output (Bolton 2002).

Regarding controllers typologies, Fig. 3.4b briefly summarizes the main options that can be selected. As mentioned before, look-up tables are often used to perform feedforward control, whereas the most widespread feedback controller corresponds to the proportional–integral–derivative (PID) one, which as it is well known can be reduced, provided the system to be controlled allows it, to the simpler PI controller (Franklin et al. 2006).

When considering model-based control, such an approach can be used both in open-loop and closed-loop configurations. The most prominent example of the first

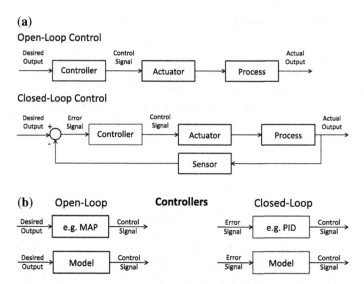

Fig. 3.4 Control configurations (**a**) and main controller types (**b**)

category is the receding horizon controller, also known in literature as model predictive control (MPC) (Allgöwer et al. 2004). Particularly, MPC computes the optimal control signals through the minimization of an assigned cost function (CF). Thus, in order to perform open-loop optimization of the future control actions, the system performance is predicted by a model of the real system. This technique is suitable for those applications, where the relationships between performance, operating constraints and system states are strongly nonlinear (Hussain 1999). On the other hand, a relevant example of feedback model-based control schemes is given by neural network controllers. Particularly, a direct inverse controller can be obtained by developing a dynamical NN (e.g., RNN), whose objective is to well capture the inverse dynamics of the plant to be controlled. Afterward, a controller can be obtained by substituting the most recent output, to be fed back to the inverse RNN, by the desired target. In this way, a feedback control behavior is intrinsically provided, as the RNN will yield on output a control variable (i.e., a selected input), which aims at tracking the assigned target (Nørgaard et al. 2000; Arsie et al. 2004).

Hereinafter, both control configurations (i.e., open-loop and closed-loop) are adopted, thus allowing to enhance model-based development of SOFC controllers at all levels, as discussed in Sect. 3.1. Particularly, a proper combination of open-loop and closed-loop approaches allows well meeting control targets, while ensuring good transient response, even when the control actions involve several input and state variables. Moreover, the coupling of feedforward and feedback controllers was proven effective in reducing disturbances impact on real-world control performance, as well as to reduce instability risks associated to feedback control actions (Bolton 2002).

3.2 Supervisory Control

Figure 3.5 particularizes the generic supervisory control level shown in Fig. 3.1 for a typical SOFC unit installation. The load demand (i.e., P_{load}) is split, in such a way as to make FCS and battery concur in meeting it. The most suitable split, which is mathematically expressed by the splitting index SI defined by Eq. (3.1), is here determined via the heuristically derived Fuzzy logic (Passino and Yurkovich 1998) map shown in Fig. 3.6a:

$$\begin{cases} SI = \frac{P_{\text{batt}}}{P_{\text{load}}} \\ P_{\text{SOFC,net}} = P_{\text{load}} \cdot (1 - SI) \end{cases} \tag{3.1}$$

As already discussed in Sect. 3.1, at this control level, the objective is to optimize load demand share between FCS and battery, mainly aiming at attaining best efficiency, while ensuring charge sustaining operation of battery pack. Therefore, the desired SI value is computed, as a function of current battery state of charge (SOC) and P_{load}, by means of the heuristic rules shown in Fig. 3.6b. Particularly, the rules, whose quantitative and qualitative descriptions are, respectively, given in Fig. 3.6a, b were defined aiming at meeting two objectives: first of all, the FCS shall be operated in its most efficient region, shown in Fig. 3.7; then, SOC variation should be maintained with a safe range around the value of 0.7, thus guaranteeing that low internal resistances occur both in charge and discharge phases (Choi et al. 2003). Therefore, when SOC is low (i.e., second column in Fig. 3.6b), the fuzzy logic membership functions yield on output negative values as long as the SOFC

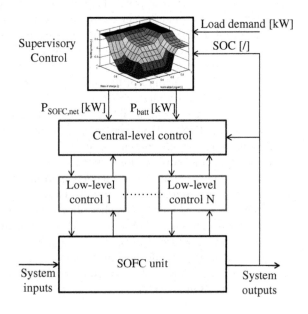

Fig. 3.5 Particularization of the generic supervisory-level (see Fig. 3.1 and Sect. 3.1) to the high-level control of SOFC units

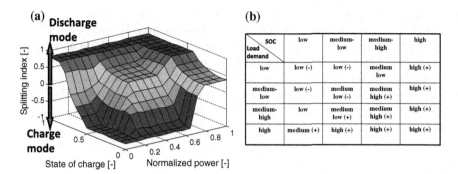

Fig. 3.6 Most suitable splitting index surface (see Eq. 3.1) **a** associated to the heuristic rules **b** determined for the SOFC unit shown in Fig. 3.2. Particularly, the table **b** shows *SI* qualitative variation as a function of load demand and battery state of charge

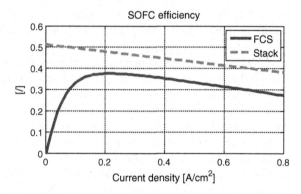

Fig. 3.7 Modeled efficiency curves for the SOFC system shown in Fig. 3.2 (Sorrentino and Pianese 2009). It is worth remarking that system efficiency is here computed on a higher heating value basis

load that meets overall power demand is lower than 0.4 A/cm^2 (see Fig. 3.7). Such a choice is directly linked to the typical behavior of FCS efficiency with respect to power load, as shown in Fig. 3.7. Indeed, the higher impact of parasitic power adsorption at low loads (Arsie et al. 2006) causes net efficiency to dramatically reduce with respect to the highest gross efficiency values, which for fuel cell are attainable exactly at those conditions (Larminie and Dicks 2003). On the other hand, at average loads, where fuel cell gross efficiency decay is still limited, the parasitic power impact decreases significantly, thus yielding the highest FCS efficiency values, as shown in Fig. 3.7. The latter aspect suggests to increase FCS power contribution in case of low load demand, using the extra power to charge the battery, as shown in Fig. 3.6. Of course, the higher the load demand (e.g., higher than 0.4 A/cm^2), the lower the FCS efficiency, due to the increasing impact of concentration losses (see Fig. 3.7). In such conditions, which are expected to be quite rare provided that a conservative sizing of SOFC stack was performed in the design phase, the best choice is to involve the battery in meeting the power demand, up to the extent allowed by current SOC.

The above considerations were extended to the other columns (i.e., corresponding to medium and high SOC values), thus yielding as a result the *SI* map shown in Fig. 3.6a. It is worth remarking here how the hierarchical structure of models described in Fig. 2.2 was here deployed first to estimate FCS efficiency variation as a function of current density (see Fig. 3.7). Then, the membership functions were tuned via a trial-and-error analysis conducted on the SOFC unit simulator associated to Fig. 3.1, whose details can be retrieved from Sorrentino and Pianese (2009).

3.3 Central-Level Control

Due to the risk of highly damaging thermal stresses induced by over-fast variation of operating condition (Aguiar et al. 2004), load-following management of SOFC systems must be carefully approached. As a matter of fact, highly fluctuating power demand applications entail selecting hybrid FCS configuration, in which a suitable transition strategy from one operating mode to another is required. As suggested in previous scientific contributions (Guezennec et al. 2003; Gaynor et al. 2008; Arsie et al. 2007), potentially damaging severe transients for FCS can be avoided by suitably limiting power slew rate. The latter was therefore adopted in the current book to develop a central-level controller. Figure 3.8 illustrates such an intermediate controller, which consists of a rate limiter, applied to the net SOFC power request made by the supervisory-level (see Fig. 3.5), and a power compensation node ensuring that transient residual power request be met by the battery pack. Particularly, Fig. 3.8 highlights how the effect played by the rate limiter during transients, which causes the FCS effective power (i.e., $P_{net,SOFC,eff}$) to be lower than the amount requested by the supervisory-level, is compensated by forcing the battery pack to meet such a residual power demand (i.e., $P_{batt,eff} \geq P_{batt}$ by the amount needed for compensating FCS slew rate).

As for the selection of the most suitable slew rate to be adopted to avoid inducing over-severe thermal transients, the SOFC simulator associated to Fig. 3.2 was deployed to perform the parametric analysis detailed in Table 3.1. Particularly, Table 3.1 lists the variation of ΔT derivative, where $\Delta T = (T_{out} - T_{in})$ is expressed as the spatial increase in temperature going from stack inlet to outlet, as a function of power slew rate. The derivative values shown in Table 3.1 were obtained by imposing a variation in SOFC net power from the most efficient point, corresponding to net AC SOFC power and current density equal to 2.31 kW and 0.25 A s^{-1}, respectively, up to the maximum FCS AC power (i.e., 5 kW, see Fig. 3.2), as shown in Fig. 3.9a.

The derivative value obtained in case no rate limiter is adopted (i.e., corresponding to an infinite slew rate) results in an unacceptable transient, shown in Fig. 3.9b, c which causes the spatial gradient across fuel cell channels to dangerously rise up to over 200 Kelvin degrees within an extremely short time window

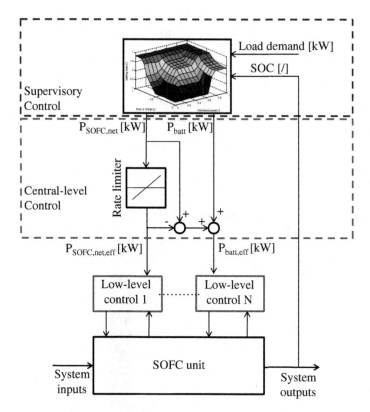

Fig. 3.8 Central-level controller for addressing safe transition from one operating mode to another in a hybridized SOFC unit

Cases	Power slew rate (W s^{-1})	Maximum ΔT derivative (K s^{-1}), $\Delta T = T_{out} - T_{in}$
1	Inf	635
2	800	50
3	400	23.5
4	80	4.75
5	8	0.5

Table 3.1 Inputs and outputs associated to the parametric analysis carried out to assess the positive influence of introducing a power rate limiter on temperature derivative

(about 3 s). On the other hand, the introduction of a finite slew rate allows limiting such a temperature rise, down to a few Kelvin degrees in correspondence of 8 W s^{-1}, as shown in Fig. 3.9c—Case 5. The latter slew rate value not only allows meeting the derivative safe limit, assumed equal to 0.5 K s^{-1} as indicated by Ferrari et al. (2005), but also enables a smooth transition from the initial to the final operating condition, as shown in Fig. 3.9b.

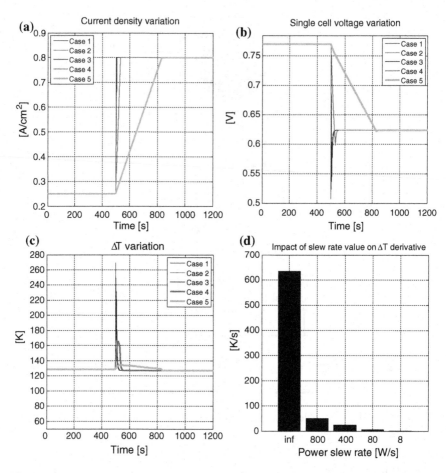

Fig. 3.9 Simulated trajectories resulting from the parametric analysis detailed in Table 3.1 (**a–c**). Bar plot (**d**) highlights the need for introducing quite a high slew rate to avoid incurring severe thermal stresses during load-following operation

3.4 Low-Level Control

Following the action performed by the central-level controller, whose main objective is to avoid over-abrupt changes in SOFC load (see Sect. 3.3), the low-level control task for a hybrid SOFC unit must be developed aiming at suitably performing energy and thermal management. The former goal is obtained by properly selecting actual operating current as a function of power demand. On the other hand, a satisfactory thermal management is attained by regulating SOFC stack outlet temperature, especially for those applications specifically destined to transportation uses (Botti et al. 2005). This is due to the severe thermal stresses imposed on cell materials by temperature variation subsequent to load change. Previous studies (Braun 2002; Aguiar et al. 2005) indicated that 100 cm^2 electroactive area planar designs require limiting temperature increase across the cell within the range 100–150 °C, to ensure cell components integrity. As discussed above, such a requirement can be safely met by suitably regulating the drive motor of the air compressor (see Fig. 3.2).

Moreover, two additional issues must be certainly taken into account: (i) the need for limiting temperature derivatives during system cold-start and shutdown; and (ii) SOFC stack does not deliver any power until its operating temperature overcomes a material-dependent threshold, here set to 700 °C (Weber and Ivers-Tiffée 2004). Therefore, two separate control problems must be faced: the first aimed at fulfilling load demand at warmed-up (i.e., regime) conditions; and the second concerning proper thermal management of the SOFC stack during cold-start phases.

As shown in Fig. 3.10, the best way to manage the transition from cold-start to warmed-up control task is to pass from one to another as a function of current outlet

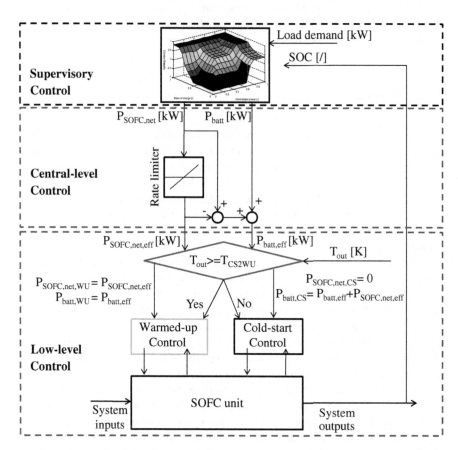

Fig. 3.10 Multilevel architecture with detailed description of low-level controllers needed to ensure proper management of the entire SOFC unit, during both warmed-up and cold-start phases

temperature. As it will be clarified later on in this section, such a transition cannot be abrupt (i.e., it cannot be a "switch"); indeed a smooth transition from cold-start (CS) to warmed-up (WU) phases must be performed to avoid causing potentially damaging thermal gradients (Sorrentino and Pianese 2011). Of course, the fact that no power can be supplied by the SOFC during most part of the CS phase has a significant impact on the SOFC unit design phase. The authors themselves showed in a previous contribution (Sorrentino and Pianese 2009) how a significantly larger battery pack shall be selected, as compared to APUs equipped with faster energy units, such as PEM fuel cells or internal combustion engines. Indeed, such a design solution allows well coping with the above-described limitations associated to long CS phases, during which battery contribution to meet load demand can be significantly increased, as shown in Fig. 3.10 (i.e., particularly referring to the variables $P_{batt,eff,WU}$ and $P_{batt,ref,CS}$).

The following two sub-sections focus on model-based development and implementation of warmed-up and cold-start low-level control tasks for hybrid SOFC units, respectively. Moreover, significant attention is paid to the interaction among control variables and key process variables, such as cold-start duration and T_{out} derivative during CS phases.

3.4.1 Warmed-up Phase

Due to the well-known limitations imposed by the relatively low tolerance of SOFC units components (i.e., both at stack and balance of plant level) to high temperature derivatives and gradients, a suitable combination of feedforward and feedback controllers is required. The authors themselves have shown, in a previous contribution (Sorrentino and Pianese 2011), that the coupling of feedforward and feedback control strategies enhances the thermal management of components characterized by several energy and flow interactions, as well as by different dynamic responses.

More in detail, Fig. 3.11 shows that the feedback control logic is applied to the air compressor, by means of the proportional–integral (PI) controller developed via model-based approach in (Sorrentino et al. 2008), in such a way so as to feed the excess air λ required to meet the desired temperature value at stack outlet. Figure 3.12a confirms that the above-mentioned PI controller guarantees that the targeted outlet temperature is reached through a suitable regulation of excess of air, as shown in Fig. 3.12b. Such a control action, imposed to the air compressor, results in a fast compensation of the thermal dynamics described in Sect. 2.3.1.2, which in turn causes voltage undershoot and relaxation time to significantly reduce and shorten, respectively (see Fig. 3.12a), as compared to open-loop response (see Fig. 2.27 and Table 2.4). It is interesting to note how the stack dynamical model well reproduces the negative impact played by temperature limitation, as imposed by the PI controller, to SOFC performance, due to polarization losses increase subsequent to outlet (and thus average) temperature decrease (see Sect. 2.2.1.8). It is also worth here mentioning

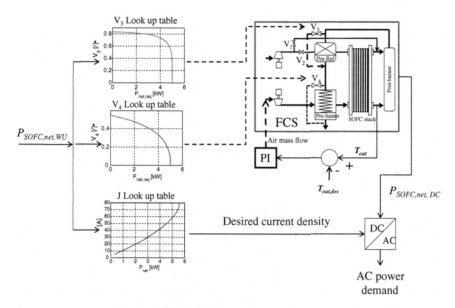

Fig. 3.11 Schematic description of warmed-up low-level controllers used for thermal and energy management of the SOFC unit shown in Fig. 3.2

that the response illustrated in Fig. 3.12 was obtained by disabling the power slew rate included in the central-level controller presented in the previous section.

On the other hand, the feedforward approach is deployed to control the heating-up action performed by pre-reformer and air preheater (see Fig. 3.10), as well as to make the SOFC stack work at the desired current. As for the former task, the upper two look-up tables, shown on the left-hand side of Fig. 3.11, are adopted to select the most suitable aperture of the bypass valves V_3 and V_4. The aim of the latter regulation is to vary the amount of hot gases flowing through pre-reformer and air preheater, in such a way as to ensure inlet anode and cathode temperatures always set close to the selected operating value (here assumed equal to 700 °C, see Fig. 3.2). A model-based trial-and-error analysis was performed in (Sorrentino and Pianese 2011) to develop these two look-up tables, which estimate, as a function of current load, the valve opening to be imposed to attain targeted hot gases mass flows, as indicated by the higher control levels (see Fig. 3.10). Tables 3.2 and 3.3, beyond showing the impact of current operating condition on most critical temperature values within FCS balance of plant, also confirm the satisfactory control action performed by the above-described V_3 and V_4 look-up tables. It is worth here reminding that during WU conditions the valves V_1 and V_2 (see Fig. 3.10) are fully closed and open, respectively.

Finally, the proper current at which the SOFC stack shall be operated to meet power demand is selected through the third look-up table, shown at the bottom left corner of Fig. 3.11. As for the valves tables, even the current versus power look-up table was developed by exploiting the FCS simulator presented in Sorrentino and

Fig. 3.12 Response of controlled variables (**a**, **b**) and control input (**c**) to a load variation from 40 to 50 A (i.e., from 0.4 to 0.5 A cm^{-2})

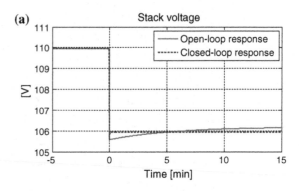

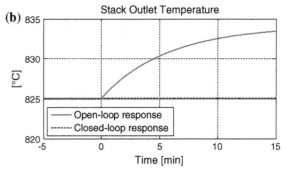

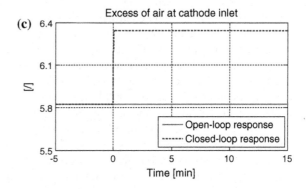

Table 3.2 Inlet and output temperatures, attained through V_3 valve management, at pre-reformer inlet and outlet (see Fig. 3.2)

Current density (A cm^{-2})	Hot fluid temperature (°C)		Cold fluid temperature (°C)	
	Inlet	Outlet	Inlet	Outlet
0.2	1061	855.2		705.1
0.5	1010	845	300	701.3
0.8	976.3	839.3		698.1

Table 3.3 Inlet and output temperatures, attained through V_4 valve management, at air preheater inlet and outlet (see Fig. 3.2)

Current density (A cm^{-2})	Hot fluid temperature (°C)		Cold fluid temperature (°C)	
	Inlet	Outlet	Inlet	Outlet
0.2	855.2	397.9		695.5
0.5	845	356.2	300	696.4
0.8	839.3	339.4		697.8

Pianese (2009, 2011). As discussed above (see Sect. 3.1.1), the attainment of the desired current level entails properly acting on the DC/AC power conditioning device, as shown at the bottom right corner of Fig. 3.11. Further details on model-based methodologies, which can be suitably deployed for addressing current control management of fuel cells, are retrievable from the contribution provided by Valdivia et al. (2014).

3.4.2 Cold-Start Phase

SOFC cold-start should be managed considering that system start-up mainly consists of three phases: cold-start, transition between cold-start and warmed-up (CS2WU), and start of warmed-up phase. In this way, a safe step-up from zero load, at ambient temperature, to warmed-up conditions can be guaranteed, thus also ensuring integrity of the heterogeneous stack materials (Ormerod 2003). As previously discussed in this book, such a key and strategic task, both in terms of system lifetime and need for limiting degradation impact, can be successfully addressed by performing model-based analyses and control rules development.

A comprehensive description of the procedure here proposed to face this operational problem is shown in Fig. 3.13. Of course, the reader must be aware that there are a number of heating-up strategies that can be adopted for SOFC stacks, as detailed in many contributions available in literature (Rancruel and von Spakovsky 2005; Apfel et al. 2006). The one presented in the following must be considered as a useful example, through which it is possible to highlight the potentialities offered by model-based approach to develop suitable controllers for addressing such a key problem in SOFC thermal management.

During the CS phase, the excess of air, differently than WU phase, is used to warm the stack up. Such a warming up effect is obtained by suitably acting on valves V_1 and V_2, which are kept open and closed, respectively, in this phase. This valves management causes the methane flow to bypass the anode channels and, thus, directly enter the postburner, as shown in Fig. 3.14. In the postburner, the methane reacts with the air coming out of the cathode channels, thus releasing the heat that will then be transferred to the incoming air fed by the compressor. Thus, the stack gets internally heated by the air flowing through the cathode channels, as it is usually required for SOFC technology (Apfel et al. 2006).

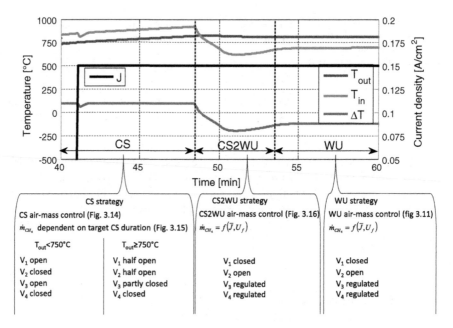

Fig. 3.13 Schematic description of model-based derived thermal management of SOFC unit during transition from cold-start to warmed-up conditions

The incoming air flow is the control variable to be managed to ensure that temperature gradient across gas channels is safely limited under 100 °C, as shown in the control scheme depicted in Fig. 3.14.

On the other hand, during the entire CS phase (see Fig. 3.13) the methane flow is kept constantly equal to a specific value, selected via the relationship methane flow-CS duration described in Fig. 3.15. Particularly, the FCS simulator presented in (Sorrentino and Pianese 2009, 2011) was exploited to run a parametric analysis, whose outcomes were suitably curve-fitted to derive the relationship shown in Fig. 3.15. The analysis of such a figure, which is of high practical relevance due to the significant impact of CS phase duration on FCS design, failure preventions, and lifetime, highlights how more transportation-/automotive-oriented applications entail increasing the amount of methane directly fed to the postburner. Moreover, the methane flow value associated to typical SOFC stationary applications (i.e., CS phase duration >5 h in Fig. 3.15), well agrees with real-world installations, as confirmed considering the experimental value shown in Fig. 3.15 (FCLAB 2014). It is finally worth noting a further benefit related to the model-based analysis illustrated in Fig. 3.15, namely the opportunity of deriving a useful correlation, also shown in the figure, to evaluate expected CS phase duration as a function of the methane mass flow directly supplied to the postburner (see Fig. 3.14).

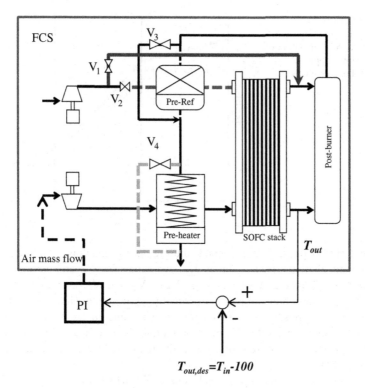

Fig. 3.14 Thermal management of the SOFC unit during cold-start

Fig. 3.15 Model-based evaluation of cold-start duration dependence on methane mass flow supplied to the postburner (see Fig. 3.14)

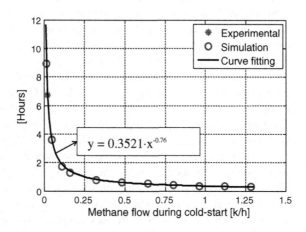

Since no current can be drawn until stack temperature becomes lower than 700 °C, as shown in Fig. 3.13, it was assumed that no load is applied before T_{out} reaches 750 °C; this means that the bulk temperature is safely higher than 700 °C whatever current is drawn out of the stack. Once such condition is reached, the end of

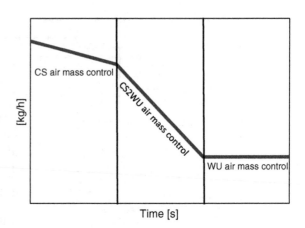

Fig. 3.16 Qualitative description of the transition from CS to WU regulation of air-mass flow supplied to SOFC cathode channels

cold-start (ECS) phase is approached, thus requiring enabling a different valves management. Particularly, as soon as T_{out} overcomes 750 °C and current can thus be drawn out from the stack (see Fig. 3.13), V_1 and V_2 have to be placed in half-open position, thus allowing to equally split the methane flow between pre-reformer and postburner. On the other hand, due to the need for enabling external pre-reforming, valve V_3 must be partially closed, thus enabling the post-burner to provide the pre-reforming with the required heat for its inner reactions (see Fig. 3.14).

Afterward, before entering the WU phase, for which the energy and thermal management strategies described in Sect. 3.4.1 shall be enabled, a proper transitioning phase, i.e., phase CS2WU in Fig. 3.13, must be included. Particularly, the incoming air flow has to be managed in such a way as to avoid abrupt and potential damaging variation in inlet air flow temperature; such a phenomenon is linked to the need for switching from a heating (in CS phase) to a cooling effect played by cathode air in the WU phase. Therefore, the authors themselves proposed in a previous contribution (Sorrentino and Pianese 2011) to replace PI-based control of air flow by a feedforward algorithm, labeled as CS2WU λ controller in Fig. 3.13. Particularly, the latter controller consists of a first look-up table, once again developed via model-based approach, that estimates the amount of air to be fed at the very beginning of the CS2WU phase, to ensure a smooth and, thus, safe variation of incoming air flow temperature, when switching from CS to WU λ control targets. Moreover, during the entire CS2WU phase, a linear variation (see Fig. 3.16) is imposed to incoming air flow, starting from the value indicated by the above-mentioned look-up table, down to the air flow requested in the subsequent WU phase. Of course, such a strategy entails developing a second look-up table, which has to correlate the WU cathode air flow (see Fig. 3.16), for a single cell, to the operating current density. As a consequence of the above observations, it shall be pointed out that SOFC start-up must be performed not only by fixing a priori the amount of methane mass flow, as required during the CS phase to attain the desired cold-start duration (see Fig. 3.15), but also by forcing the FCS to operate at a given current density in the very beginning of the WU phase. Otherwise, the most suitable smooth variation of air-mass flow, as described in Fig. 3.13, could not be enabled. It is worth noting in Fig. 3.13 how the transition from CS to WU phase causes ΔT across stack channels to vary from positive to negative values, as a consequence of the different control targets to be met in these two distinct phases (see Figs. 3.11 and 3.14).

It is finally worth here remarking how a strategy similar to the one presented above must be developed for system shutdown, this time aiming at managing inlet mass flows in such a way as to ensure that negative temperature derivative does not exceed the targeted safe value. Of course, system shutdown can be considered less-dependent on SOFC unit end application and, thus could be less critical: for instance, even for transportation applications, shutdown duration can be longer that cold-start, especially if the entire nighttime is available.

3.5 Case Studies

In the following sub-sections, two real-world applications of model-based energy and thermal management of SOFC units are proposed. The former one corresponds to the installation of a hybridized SOFC unit on a heavy-duty truck, aimed at replacing the internal combustion engine (ICE) during hotel operation (Lutsey et al. 2004). The latter case study consists in a CHP application for a typical European dwelling.

The above-analyzed cases serve as a useful reference for fuel cell practitioners, not only for well distinguishing between different scopes and energy-economic benefits associated to the introduction of SOFC-based energy plants, but also to emphasize the need for adapting energy and thermal management and, as a consequence, material specification requirements, to the specific end application.

3.5.1 Case Study 1: Automotive APU

The hybrid SOFC-APU shown in Fig. 3.2 is suitable to meet the power demand of a heavy-duty truck during hotel operation, especially at nighttime when the driver stops along his route. The adoption of a fuel cell for such an automotive use is particularly effective, due to the opportunity of replacing the ICE when hotel power is requested, for instance to operate fans, heaters, air-conditioning system, refrigerator, and so on (Lutsey et al. 2004). Indeed, currently these power demands are satisfied by operating the ICE the truck is equipped with to guarantee traction, in idle conditions, thus causing significant fuel consumption increase as well as undesired engine emissions. Thus the high efficiency guaranteed by an SOFC-APU makes such a solution very attractive to avoid the above-mentioned issues associated to engine idling.

In this section, it is proposed to meet the typical power trajectory associated to hotel operation on a heavy-duty truck, shown in Fig. 3.18a, by a hybrid SOFC unit capable of delivering 2.31 kW nominal electric power at 38 % efficiency (see Fig. 3.2). Of course, thermal management of such an SOFC unit must be carefully addressed, since its start-up time is expected to be significantly longer than ICE. Therefore, the control and thermal management rules presented in Sect. 3.4 were deployed for two main purposes: (i) the amount of methane to be supplied to the postburner to obtain the desired start-up time is found by appropriately referring to Fig. 3.15 and (ii) the proper energy and thermal management during warmed-up conditions, here corresponding to the time window in which the hotel operation is active, was guaranteed by deploying the entire multilevel architecture described throughout this chapter.

Table 3.4 summarizes main data and specifications of the battery pack, which the hybrid SOFC-APU is equipped with. Particularly, the number of cells was selected by guaranteeing that the minimum state of charge (SOC) never falls below a safe limit, here set to 25 %, even when the driver cannot predict in advance when he has to stop and, thus, the SOFC start-up time coincides with hotel operation beginning. Preliminary results were already presented by the authors in a previous contribution (Sorrentino and Pianese 2009), where the model-based selection of both battery cells and FCS system components is treated in detail.

As for battery pack modeling, namely the method used to estimate current SOC as a function of current power, an equivalent circuit model is adopted, consisting of a voltage source and a resistance in series. Figure 3.17 shows the model particularized to the two possible operating modes, i.e., charging and recharging. The dependence of open-circuit voltage and internal resistance on SOC is accounted for by referring to the look-up tables proposed in (Markel et al. 2002). Then, actual SOC is computed solving the following nonlinear dynamic equation for a single cell:

$$\frac{dSOC}{dt} = -\frac{I_c}{Q_{\max}} = -\frac{V_0(SOC) - \sqrt{V_0(SOC)^2 - 4R_{in}(SOC)P_{Batt}/N_B}}{2R_{in}(SOC)Q_{\max}} \quad (3.2)$$

The latter approach, which is certainly simplistic when detailed studies on battery performance and state of health are to be conducted, was proven particularly effective when applied to energetic evaluations involving batteries (Sundström et al. 2010). Further details on the mathematical approach to be followed to correlate SOC to current battery power (see Eq. 3.2) are retrievable from Rizzo et al. (2014).

Table 3.4 Battery pack specifications (Johnson 2002)	Type	Lead–acid
	Nominal cell voltage (V)	12
	Cell capacity (Ah)	25
	Number of cells	15
	Connection	Series

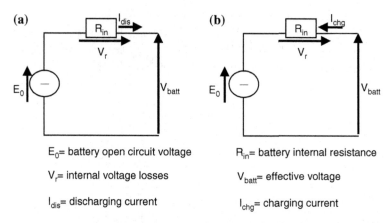

E_0= battery open circuit voltage R_{in}= battery internal resistance

V_r= internal voltage losses V_{batt}= effective voltage

I_{dis}= discharging current I_{chg}= charging current

Fig. 3.17 Modeling approach adopted to simulate battery pack performance: **a** discharge mode; **b** charge mode

Figures 3.18 and 3.19 show the results yielded on output by the simulation of the hotel power trajectory shown in Fig. 3.19a. Particularly, Fig. 3.19a shows that the SOFC unit was turned on about 2 h before the effective hotel power request beginning, thus allowing not to over-deplete the battery pack (see Fig. 3.18c) during FCS CS phase, when only negative net power is obtained due to air compressor absorption in the SOFC system (see Sect. 2.3.3). Afterward, the supervisory-level controller (see Fig. 3.5) was deployed to appropriately split the power demand between SOFC unit and battery pack (see Fig. 3.19a), which in turn led to obtain an overall charge sustaining operation for the battery pack, as shown in Fig. 3.19c. On the other hand, Fig. 3.21 clearly highlights how the proposed supervisory-level, even after the subsequent actions performed by the central- and low-level controllers, ensures operating the SOFC unit mostly within its best efficiency domain.

Furthermore, Fig. 3.19 highlights how the power slew rate introduced by the central-level controller (see Sect. 3.3) allowed limiting temperature excursions induced by load-following operation in all main SOFC unit components. Such a strategy guarantees material integrity and, in turn, allows increasing FCS lifetime to the maximum extent.

The deployment of the low-level controller described in Sect. 3.5 allowed to properly perform thermal management throughout FCS operation, particularly ensuring a safe transition from cold-start to warmed-up phases. Moreover, the methane mass flow sent to the postburner during the CS phase, selected via the relationship described in Fig. 3.15 and depicted in Fig. 3.18d, was demonstrated effective to achieve the targeted start-up time (i.e., less than 2 h, as shown in Fig. 3.18b).

Finally, a summary of the main energy benefits achievable by introducing hybrid SOFC units in the automotive sector is highlighted in Table 3.5. Particularly, it emerges how the SOFC methane consumption is much lower, by approximately

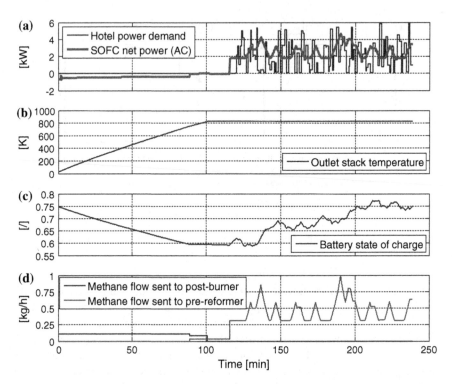

Fig. 3.18 Transient behavior of main performance and control variables yielded on output by the first case study simulation: **a** Power demand and net SOFC power (AC) trajectories; **b** Cold-start (i.e., until 100 min) and warmed-up SOFC outlet temperature; **c** Battery state of charge evolution during SOFC cold-start and warmed-up; **d** Trajectories of methane flow entering postburner (for cold-start purposes) and pre-reformer

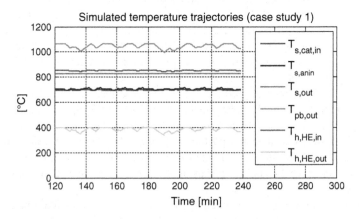

Fig. 3.19 Temperature trajectories simulated in SOFC warmed-up conditions

Table 3.5 Case study 1: summary of main results

	SOFC-APU	ICE idling
Methane consumption (kg)	1.15	2.71[a]
Useful heating energy delivered on output (kWh)	4.8	0[b]

[a]Equivalent methane mass consumed by ICE idling is evaluated as $m_{CH4,ICE,idle} = m_{Diesel,ICE,idle} \cdot HHV_{Diesel} \cdot HHV_{CH_4}^{-1}$, where diesel consumption is estimated assuming a 10 % ICE idling efficiency
[b]Of course, thermal power could be recovered from the thermal engine as well, at the cost associated to required powertrain modifications (e.g., inclusion of further piping elements, heat exchangers, and so on)

60 %, than the equivalent methane consumption associated to a diesel ICE, which is operated at idle speed to meet the same hotel power trajectory. Moreover, the availability of an SOFC-APU inside truck cabin provides a further benefit, namely the generation of useful thermal power, to be potentially used to meet cabin heating demand.

3.5.2 Case Study 2: Residential CHP

The second case study concerns the deployment of the same SOFC unit, as the one embedded in the hybrid APU shown in Fig. 3.2, as a combined heat and power plant (CHP) aimed at meeting the daily thermal demand of an exemplary European dwelling (Brandon 2014), as shown in Fig. 3.20. The heating demand trajectory, shown in the latter figure, was obtained by randomly generating a thermal power demand vector covering a yearly averaged daily timeframe, with power ranges that

Fig. 3.20 Temperature trajectories simulated in SOFC warmed-up conditions

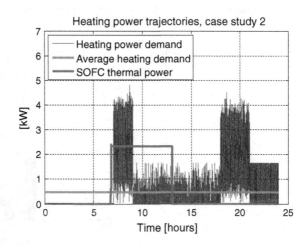

vary along the daytime. Particularly, the maximum range value was imposed in the morning (i.e., between 7 and 9 am), when a lot of thermal power is required for sanitary water heating, and in the evening (i.e., between 6 and 9 pm), when thermal power is required for room heating (in winter time) and other uses (e.g., cooking, dishwasher, etc.).

As for the operating point selection, it was chosen to let the SOFC CHP operate at its nominal power (i.e., most efficient FCS operation), corresponding to 2.31 kW AC power (see Figs. 3.2 and 3.21). In this condition, the SOFC unit is also capable of yielding on output as high as 2.34 kW thermal power, in accordance with typical SOFC heat-to-power ratio (Brandon 2014). Therefore, in order to meet the fluctuating heating demand, a thermal storage system (TES) must be coupled to the SOFC unit. Upon availability of a TES, the SOFC operating time $t_{SOFC,CHP}$, needed to guarantee requested heat being met throughout the day, can be evaluated as a function of average heat power demand (see Fig. 3.20) and thermal storage efficiency, as follows:

$$t_{SOFC,CHP} = \frac{\overline{P}_{heat,dwell} \cdot 24\,[h]}{\eta_{TES} \cdot P_{heat,SOFC}\,|_{P_{net,SOFC,AC}=2.31\,kW}} = \frac{0.4855\,[kW] \cdot 24\,[h]}{0.8 \cdot 2.34\,[kW]} = 6.22\,[h]$$

$$(3.3)$$

In Eq. (3.3) the thermal storage efficiency η_{TES}, defined as the ratio between used thermal energy and stored one, was safely set to 0.8, in accordance with what was proposed by Ataer (2006). It is worth noting that current technology would suggest not to perform daily cycles of start and shutdown of residential SOFC systems, so as not to accelerate thermal stress-related degradation. On the other hand, over the entire lifetime, the possibility for necessarily stopping the operation should be accounted for, e.g., system maintenance. Moreover, stopping system operation allows modifying original energy management strategies, in such a way as to adapt them to degrade but still effective functioning of the SOFC unit. In these cases, the

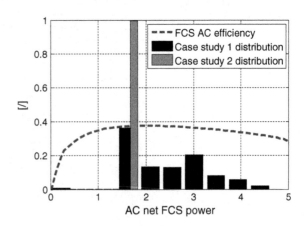

Fig. 3.21 Operating domains simulated in case study 1 and case study 2

Table 3.6 Case study 2: summary of the main results

	Methane cost (Eur year^{-1})	Electricity savings (Eur year^{-1})	Annual savings (Eur year^{-1})
Conventional (i.e., boiler only)	A = 371.6	0	0
SOFC CHP	B = 1011.9	C = 888.6	AS = A–(B–C) = 248.3

proposed model-based evaluation of the adopted start-up procedure can be useful to meet household load demand, as well as comply with start-up time constraints. Once the warmed-up duration had been found, the next step was to select the methane flow to be supplied to the postburner to ensure useful heating be delivered at the desired time. As it is shown in Fig. 3.20, the first peak in heating power demand occurs on average around 7 am. Therefore, it was selected, from Fig. 3.15, the methane flow ensuring the SOFC CS phase be over before 7 am. Particularly, a value of 0.05 kg/h of methane results in 3.7 h CS duration, thus allowing to safely set the automatic SOFC turn-on around 4 am, as shown in Fig. 3.20. Such a choice is preferable since it allows reducing the recourse to thermal storage as much as possible, which in turn makes the time estimation performed via Eq. (3.3) more reliable.

Table 3.6 synthesizes the results obtained in terms of annual savings, estimated as a function of Italian[1] electricity and methane costs, assumed equal to 0.23 and 0.083 Eur kWh^{-1}, respectively. It is worth pointing here out that annual savings were estimated as the difference between conventional system (i.e., without SOFC CHP) methane cost, estimated assuming that the heat power demand shown in Fig. 3.20 is met by a 90 % efficiency boiler (Staffel et al. 2008), and the SOFC operating cost. The latter variable was estimated as the difference between the cost of methane, used during both CS and WU phases, minus the electricity savings obtained by supplying the SOFC AC power either to the dwelling load or to the grid.

The annual savings estimated through the simulation discussed above are of a certain relevance, since, if the projections on SOFC investment cost will be met (i.e., about 300 Eur/kW in a near future, (Staffel et al. 2008)), they will result in a simple payback as low as 2.8 years.

Finally, also for case study 2 the operating point distribution over the efficiency vs power domain is illustrated in Fig. 3.21. As a consequence of the much easier energy management adopted in this case, the SOFC stack was always operated in correspondence of its nominal condition during the entire WU phase.

---

[1]Source: http://epp.eurostat.ec.europa.eu/statistics_explained/index.php/Electricity_and_natural_gas_price_statistics.

3.6   Chapter Closure

This chapter presented and discussed the deployment of the hierarchical structure of models, described and explained in Chap. 2, for performing model-based development of control strategies. Particularly, a multilevel control structure was conceived, in such a way as to allow the proper combination and interaction of different control approaches, consisting of both feedforward and feedback controllers, as well as the accomplishment of different targets.

Indeed, supervisory-level goal is to optimize power contribution split between the SOFC unit itself and either a hybridizing component (e.g., a battery pack) or the electric grid itself. To this aim, a rule-based controller was developed to optimize the above-described energy management.

Downward, the targeted set point addressed by the supervisory-level is processed by a central-level controller, whose aim is to adapt power demand to the specific requirements of SOFC transient operation, by taking advantage of the availability of the battery pack (or the electric grid) to meet abrupt changes in load request.

Finally, the low-level controller processes the information passed by the superior levels, in such a way as to achieve desired set points, while guaranteeing proper thermal and energy management of main SOFC unit components in both cold-start and warmed-up operating conditions.

The above control structure, whose suitability was verified via both numerical tests, supported by literature, and referring to relevant experimental information, was then proposed to perform energy and thermal management in two highly representative case studies. The former one corresponds to an automotive APU for supplying hotel power demand to a heavy-duty truck cabin; the latter consisted of a SOFC CHP plant, designed to provide, with the support of a thermal energy storage system, the average daily heat requested by a representative European dwelling. The results confirmed the high potential offered by methane-fed SOFC units, both to attain significant energy savings as well as to provide interesting investment revenues, especially assuming projected cost per kW targets are met in a near future.

References

Aguiar P, Adjiman CS, Brandon NP (2004) Anode-supported intermediate temperature direct internal reforming solid oxide fuel cell. I: model-based steady-state performance. J Power Sour 138:120–136

Aguiar P, Adjiman CS, Brandon NP (2005) Anode-supported intermediate-temperature direct internal reforming solid oxide fuel cell: II. Model-based dynamic performance and control. J Power Sour 147:136–147

Allgöwer F, Findeisen R, Nagy ZK (2004) Nonlinear Model Predictive Control: From Theory to Application. J Chin Inst Chem Eng 35:299–315

Apfel H, Rzepka M, Tua H, Stimming U (2006) Thermal start-up behaviour and thermal management of SOFC's. J Power Sour 154:370–378

Arsie I, Di Domenico A, Pappalardo L, Pianese C, Sorrentino M (2006) Steady-State Analysis and Energetic Comparison of Air Compressors for PEM Fuel Cell Systems. In: Proceedings of 4th ASME international conference on fuel cell science, engineering and technology, Irvine (CA), USA

Arsie I, Di_Domenico A, Pianese C, Sorrentino M (2007) Modeling and Analysis of Transient Behavior of Polymer Electrolyte Membrane Fuel Cell Hybrid Vehicles. ASME Trans J Fuel Cell Sci Technol 4:261–271

Arsie I, Pianese C, Sorrentino M (2004) Nonlinear recurrent neural networks for air fuel ratio control in SI engines. SAE Technical Paper 2004-01-1364, doi:10.4271/2004-01-1364

Ataer OE (2006) Storage of thermal energy, in energy storage systems. In Gogus YA (ed) Encyclopedia of life support systems (EOLSS), Developed under the auspices of the UNESCO, Eolss Publishers, Oxford

Bolton W (2002) Control system. First Edition Elsevier Ltd

Botti JJ, Grieve MJ, MacBain JA (2005) Electric vehicle range extension using an SOFC APU. SAE Technical Paper 2005-01-1172. doi:10.4271/2005-01-1172

Brandon N (2014) SOFCs for power plants—current status and future perspectives. In: 1st Symposium "solid oxide fuel cells for next generation power plants", June 23rd 2011, Delft University of Technology, The Netherlands. http://dutw1479.wbmt.tudelft.nl/~sofcpowergen2011/Presentations/SOFC%20for%20power%20plant.pdf

Braun RJ (2002) Optimal design and operation of solid oxide fuel cell systems for small-scale stationary applications. Ph.D. Thesis, University of Wisconsin, Madison, WI

Choi TY, Guezennec YG, Rizzoni G (2003) Supervisory control of a fuel cell SUV hybridized with supercapacitors. In: Proceedings of 4th international conference on control and diagnostic in automotive applications, June 18–20, Sestri Levante, Italy

FCLAB, Fuel Cell Laboratory University of Perugia (2014) www.fclab.unipg.it/download.html

Ferrari ML, Traverso A, Magistri L, Massardo AF (2005) Influence of the anodic recirculation transient behavior on the SOFC hybrid system performance. J Power Sour 149:22–32

Franklin G, Powell J, Emami-Naeini A (2006) Feedback control of dynamic systems, 5th edn. Prentice Hall, New Jersey

Gaynor R, Mueller F, Jabbari F, Brouwer J (2008) On control concepts to prevent fuel starvation in solid oxide fuel cells. J Power Sour 180:330–342

Guezennec Y, Choi TY, Paganelli G, Rizzoni G (2003) Supervisory control of fuel cell vehicles and its link to overall system efficiency and low-level control requirements. In: Proceedings of 2003 American control conference, June 4–6, vol 3, pp 2055–2061. Denver, CO (USA)

Hussain MA (1999) Review of the applications of neural networks in chemical process control simulation and online implementation. Artif Intell Eng 13:55–68

Johnson VH (2002) Battery performance models in ADVISOR. J Power Sour 110:321–329

Larminie J, Dicks A (2003) Fuel cell systems explained. Wiley, Chichester

Lutsey N, Wallace J, Brodrick CJ, Dwyer HA, Sperling D (2004) Modeling stationary power for heavy-duty trucks: engine idling vs. fuel cell APUs. SAE Technical paper 2004-01-1479. doi:10.4271/2004-01-1479

Markel T, Brooker A, Hendricks T, Johnson V, Kelly K, Kramer B, O'Keefe M, Sprik S, Wipke K (2002) ADVISOR: a systems analysis tool for advanced vehicle modeling. J Power Sour 110:255–266

Nørgaard M, Ravn O, Poulsen NL, Hansen LK (2000) Neural networks for modelling and control of dynamic systems. Springer, London

Ormerod RM (2003) Solid oxide fuel cells. Chem Soc Rev 32:17–28

Passino KM, Yurkovich S (1998) Fuzzy control. Addison Wesley Longman, Menlo Park

Rancruel D, von Spakovsky M (2005) Investigation of the start-up strategy for a solid oxide fuel cell based auxiliary power unit under transient conditions. Int J Thermodyn 8:103–113

Rizzo G, Sorrentino M, Arsie I (2014) Numerical analysis of the benefits achievable by after-market mild hybridization of conventional cars. Int J Powertrains (in press)

Sorrentino M, Pianese C (2009) Control oriented modeling of solid oxide fuel cell auxiliary power unit for transportation applications. ASME Trans J Fuel Cell Sci Technol 6:041011–04101112

Sorrentino M, Pianese C (2011) Model-based development of low-level control strategies for transient operation of solid oxide fuel cell systems. J Power Sour 196:9036–9045

Sorrentino M, Pianese C, Guezennec YG (2008) A hierarchical modeling approach to the simulation and control of planar solid oxide fuel cells. J Power Sour 180:380–392

Staffell I, Green R, Kendall K (2008) Cost targets for domestic fuel cell CHP. J Power Sour 181:331–349

Sundström O, Guzzella L, Soltic P (2010) Torque-assist hybrid electric powertrain sizing: from optimal control towards a sizing law. IEEE Trans Control Syst Technol 18:837–849

Valdivia V, Barrado A, Lázaro A, Sanz M, del Moral DL, Raga C (2014) Black-box behavioral modeling and identification of DC–DC converters with input current control for fuel cell power conditioning. IEEE Trans Industr Electron 61:1891–1903

Weber A, Ivers-Tiffée E (2004) Materials and concepts for solid oxide fuel cells (SOFCs) in stationary and mobile applications. J Power Sour 127:273–283

Weber A, Sauer B, Muller A, Herbstritt D, Ivers-Tiffée E (2002) Oxidation of H_2, CO and methane in SOFCs with Ni/YSZ-cermet anodes. Solid State Ionics 152:543–550

Chapter 4
Models for Diagnostic Applications

4.1 Model-Based Diagnosis

Fault detection and isolation (FDI) issues have been investigated, since the 1970s, in parallel with the increase in system automation degree (Simani et al. 2003). The main reason leading to this research field was the need for cheaper and more reliable microcomputers, related to sensors and actuators production (Isermann 2006). These studies initially dealt with the investigation of several approaches, driving the attention from traditional physical redundancy to more advanced analytical redundancy methods. These latter show quite different features from the former ones (Simani et al. 2003): on the one hand, physical redundancy methods exploit physical devices equipped onboard to replace a regular component in case of fault occurrence; on the other hand, analytical redundancy methods use mathematical models, based on physical equations or signal features, to simulate the process behavior. In this latter case, the detection and isolation of malfunctions are performed by treating the data extracted from the models, with a few advantages with respect to the physical redundancy methods. Indeed, the use of models allows avoiding the use of additional equipment, with a consequent reduction in system hardware costs. Nevertheless, high reliability and accuracy of the models is strongly required.

Generally speaking, fault diagnosis is mainly based on three tasks (Isermann 2006; Simani et al. 2003; Witczak 2003): (i) fault detection, (ii) fault isolation, and (iii) fault identification. The aim of the *fault detection* task is to detect the presence of an abnormal system status (i.e., a fault or a malfunction), which is not coherent with the expected normal behavior related to the current operating condition. *Fault isolation* follows the detection and its purpose is to define the fault type and its spatial and time location (i.e., which component(s) experienced the fault). The last task is known as *fault identification*, through which the fault size and time-variant behavior are estimated. Usually, *fault detection* is associated to a former task, *system monitoring*, through which the main system variables are monitored and all

© Springer-Verlag London 2016

D. Marra et al., *Models for Solid Oxide Fuel Cell Systems,*
Green Energy and Technology, DOI 10.1007/978-1-4471-5658-1_4

the required information to perform a reliable diagnosis are gathered. The reliability and accuracy of the entire diagnosis process strongly depends on the approach followed during its design. First of all, a detailed study of the system configuration is mandatory to define a clear picture of which could be the most likely faults or malfunction the system can be affected by. Then, a detailed analysis of the components interaction helps understanding the plausible system response to a specific faulty state (Polverino et al. 2015).

Before going into the details of diagnostic algorithm development, more clarity should be given upon the terminology adopted in this research field, so as to distinguish among the different kinds of abnormal states a generic system can experience. Great efforts have been made in defining standard definitions to be assumed by experts in different technological areas, like, for instance, in the reliability, availability and maintainability (RAM) dictionary (Omdahl 1998). Although the terms malfunction, fault, and failure are often confused, each one expresses a specific and well-defined event. According to many renowned authors (Isermann 2006; Simani et al. 2003; Witczak 2003), a malfunction consists in an intermitted irregularity of the system operation, a fault represents a deviation of at least one system feature from its normal behavior, whereas, a failure is the permanent interruption of system performance. According to the proposed definitions, the term *fault diagnosis* denotes the capability to determine type, location, size, and occurrence of an unexpected or undesired deviation of a specific system component from its acceptable conditions.

The achievement of the aforementioned tasks is fulfilled by extracting specific features from the data acquired on the system during its operation. The features extraction procedure varies according to the considered methodology. In the available literature, several methodologies are often taken into account to cope with this task, known as model-based, signal-based, or knowledge-based approaches. A model-based approach exploits a mathematical model (i.e., state space, multidimensional, lumped, neural network, etc.) to simulate the system variables behavior during normal operations; the required features are extracted by comparing the variables measured on the system with those simulated by the model. According to a signal-based approach, the feature extraction originates from the direct evaluation of the signals acquired on the system during its operation. Differently from the former methods, a knowledge-based approach is mainly based on the encoding of the heuristic knowledge hold by human experts. These features might lead to the conclusion that the development of both signal- and knowledge-based algorithms could be faster than that of a model-based one. However, it is worth bearing in mind that signal- and knowledge-based approaches require complex experiments to be carried out both in normal and faulty conditions, in order to link the acquired information to specific faulty states (Polverino et al. 2015). From time to time, these experiments might not be performed due to technical complexity or the unfeasibility of reproducing a fault in a controlled and reproducible way (Polverino et al. 2016). Consequently, it is clear that the development of a signal- or knowledge-based diagnostic algorithm can be hindered by experiments feasibility, costs, and time issues (Polverino et al. 2015). These drawbacks headed the interest toward model-based approaches, which, despite model design complexity, show a remarkable advantage: the use of physical models reduces the need for experimental data, and, accordingly, the higher physical content contributes to increasing the generalizability of the algorithm.

4.2 Residual Generation

As previously stated, a model-based algorithm requires the development of a mathematical model to simulate the behavior of the system under investigation. Therefore, the reliability of the algorithm is related to the accuracy of the model. Despite the need for a preliminary a priori knowledge of the system, the design of the model structure can be defined following general physical laws, whose validation can be performed with an amount of experimental data determined by the model type (i.e., white-, gray-, or black-box, as shown in Fig. 2.1). To overcome the issues related to experimental activities, gray-box models may be chosen. Indeed, performing virtual experiments allows gathering a high amount of data with a sensible reduction of time and costs, for both model training and test. For diagnosis purposes, the choice of a gray-box model helps in analyzing those operating conditions complicated or even impossible to be safely reproduced on a test bench. From the assessments stated above, the mathematical model is proved being the core of the methodology and it should be capable of simulating both direct and indirect correlations among system components to perform a reliable diagnosis (Polverino et al. 2015).

When specific details concerning the system and the physical phenomena under investigation are missing or hard to retrieve, the use of data-driven models can represent a valid choice. An example of data-driven models is given by numerical maps simulating system normal operation at different operating conditions. For each operating condition, the data related to the main system variables (or to those of interest for the diagnosis) are gathered and treated in order to extract significant features (e.g., mean value, standard deviation, etc.). These features are then collected into maps or lookup tables, function of the input variable set points at the considered operating condition, as illustrated in Fig. 4.1. The pros of this approach mainly consist in the easiness in the maps development, since no geometrical feature or specific physical parameter is required, and in its fast implementation for online uses. However, the main cons resides in the need for an exhaustive experimental data set, which has to cover all the possible operating conditions the system can work in.

In the scheme proposed in Fig. 4.2, the main diagnostic tasks previously mentioned are remarked. During the monitoring process, all the variables monitored on the real system (Y), affected by noise (N), are compared to those simulated by the process model ($\hat{Y}$), accounting for the same input parameters (U). This comparison leads to the generation of specific features, known as *residuals*. In the literature, many authors define a residual as the difference between the measured and

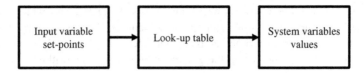

Fig. 4.1 Map-based approach for system variables simulation

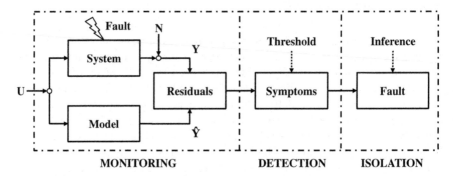

Fig. 4.2 Model-based fault diagnosis scheme, adapted from Isermann (2006), Polverino et al. (2015)

simulated signals (Arsie et al. 2010; Escobet et al. 2009; Isermann 2006; Simani et al. 2003; Witczak 2003):

$$r = Y - \hat{Y} \tag{4.1}$$

where the accounted variables can be expressed either as scalars or vectors. If several residuals are evaluated with respect to different measurements (e.g., mass flow, temperature, pressure, etc.), their physical dimensions will be different from one another. Thus, to have a uniform evaluation of the residuals and to allow handling a nondimensional vector, Eq. (4.1) can be written in the form of percentage residuals as follows:

$$r = \frac{Y - \hat{Y}}{\hat{Y}} 100 \tag{4.2}$$

As previously stressed, Eq. (4.1), as well as Eq. (4.2), expresses the deviation of the simulated parameters from those that are monitored directly on the systems through dedicated measurement devices. From a theoretical point of view, if a system is behaving normally and the model allows a correct evaluation of the system behavior, any computed residual should be equal to zero. However, when dealing with real systems, any kind of measurement is affected by noise and disturbances; moreover, although a model can be considered as much accurate as possible, there will always be uncertainties related to model parameters. Those aspects lead to nonzero residuals even during normal operation.

The above-mentioned remarks highlight the need for introducing a tolerance range in which a residual can lay during normal operation. On the one hand, if the considered residual falls within this range, it can be stated with a certain probability that the system is behaving normally; on the other hand, if the residual overcomes the set boundaries, it can be asserted that a faulty state might be occurring. However, the choice of the optimal tolerance range, and thus the related boundary thresholds, is a critical task, which has to take into account several aspects, as explained in the following section.

4.3 Threshold Design

The design of proper threshold levels for the fault detection task should take into account several aspects, from model parameters uncertainty (i.e., model inaccuracy) to measurement disturbances and signal noise. The simplest design approach for threshold levels definition is characterized by setting a specific value kept constant during system operation (i.e., fixed threshold approach). However, such an approach is not suitable for complex systems which usually runs in several operating conditions and when dealing with low-precision measurement devices. More advanced approaches (based on, e.g., statistics, fuzzy logic, etc.) can be implemented to achieve a more robust threshold design (Lo et al. 2004; Simani et al. 2003). These approaches can improve the diagnostic capability when operating with measurements showing poor resolution due to, e.g., cheap instruments or low sensitivity. In such a case, setting low thresholds may induce residuals to frequently overcome them, resulting in an almost continuous fault detection. However, the incidence of real faults might be lower than the obtained detection rate, meaning that most of the detected events are related to false alarms. Generally speaking, a *false alarm* can be defined as the probability of erroneously detecting a faulty state during a normal system operation. As a rule, when developing a fault diagnosis algorithm, the probability of false alarms should be kept as low as possible. A conceivable design path may consist in assuming high threshold levels, thus increasing the algorithm tolerance to residual deviation. However, this choice may lead to an opposite effect, i.e., missing a real fault occurring in the system. As defined for the false alarm, a *miss detection* can be defined as the probability of erroneously recognizing a real fault as normal state. The probability of false alarm and missed fault are known in the literature (especially concerning statistics and probabilistic approaches) as Type I and Type II errors, respectively. Accordingly, a summary of all the possible events that can occur when performing an inference on a system state can be resumed in Table 4.1.

What should be also taken into account is the need in certain applications for the capability of early detection (Kimmich et al. 2005). The term early detection is here assumed with respect to incipient faults, i.e., faults with low magnitude. This feature is particularly important for applications where a sudden counteraction is

Table 4.1 Table of error types (Montgomery et al. 2004)

		Inference	
		Faulty state	Normal state
Event	Normal state	Type I error False positive	Correct inference True positive
	Faulty state	Correct inference True negative	Type II error False negative

required, so as to keep a continuous system operation, or at least to drive the system into a safer operating condition, thus avoiding a possible failure. Indeed, prompt counteractions can drive the system toward a new operating condition, to prevent, for instance, system shut down if maintenance is needed, thus reducing inefficiency and maintenance costs. For this reason, the threshold levels should be set as low as possible to be able to detect faults with low magnitude.

An example of how the threshold choice can affect the inference on a residual behavior is given in Fig. 4.3. In this figure, the time behavior of a generic residual is depicted as a continuous line and two different tolerance ranges are represented with dashed and dash-dotted lines, respectively. The first tolerance range is defined as [−threshold I, +threshold I], whereas the second one as [−threshold II, +threshold II]. On the one hand, if the first range is considered, the depicted residual behavior goes beyond the chosen limits three times, inducing the detection of the same number of abnormal events (i.e., events A, B, and C). On the other hand, if the second range is assumed, only the event B can be associated to an irregular state. By analyzing the proposed example, the event B may be associated with a certain probability to a real faulty state, since the residual behavior shows a monotonous increase and overcomes both the considered threshold ranges. The other two events (i.e., event A and C) may be either generated by an incipient fault or by signal noise or disturbances, thus generating possible false alarms. From the proposed example it becomes clear how, with the same residual time behavior, different results can be obtained by varying the tolerance range.

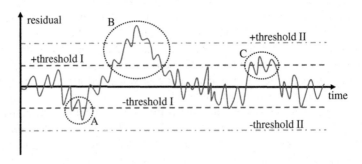

Fig. 4.3 Example of change in detection results according to different tolerance range choice

It is also worth noting that the tolerance ranges presented in the above example are defined through deterministic threshold levels. However, the variables measured on a real system generally show probabilistic features rather than deterministic ones. For instance, in steady-state conditions, the measured value of a generic variable may oscillate around its local mean value and can be represented by statistical indicators (Arsie et al. 2000). In some cases, the statistical distribution of the values can be represented by a *probability density function*, through which it is possible to define the false alarm and missed fault probabilities, once a threshold level is set. In Fig. 4.4 a comparison between deterministic and probabilistic residual evaluation is presented.

On the one hand, when assuming deterministic residuals, because no probability density function is considered either in normal or faulty conditions, the probability of missed fault or false alarm cannot be computed. On the other hand, assuming for each residual a stochastic behavior, the two probabilities can be defined as follows: the probability of missed fault is computed by intersecting the probability density function of the residual in faulty state with the threshold value (i.e., green dashed area in Fig. 4.4), whereas the probability of false alarm can be evaluated by intersecting the probability density function of the residual in normal state with the aforementioned threshold (i.e., red dashed area in Fig. 4.4) (Polverino et al. 2015).

Another important feature of foremost interest, especially when dealing with dynamic systems, is related to the ability of the diagnostic algorithm to distinguish between a residual deviation due to an undesired event and that caused by a transient maneuver (e.g., when changing operating condition). When using a steady-state model for the representation of the system behavior at specific operating conditions, any deviation from a stationary condition to another one can be

Fig. 4.4 Comparison between deterministic (*upper*) and probabilistic (*lower*) residual evaluation process, according to false alarm and missed fault probability (Polverino et al. 2015)

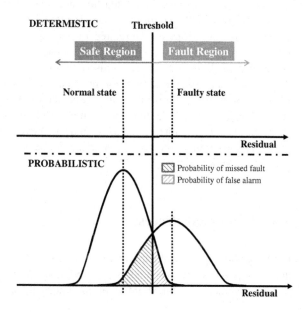

Fig. 4.5 Example of transient
maneuver not distinguishable
from a faulty event with a
diagnostic algorithm based on
a steady-state model

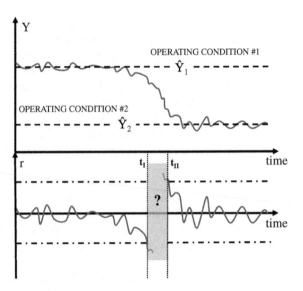

interpreted as an abnormal deviation from the normal system state. An example of
such a case is given in Fig. 4.5. At the top of the figure, the time behavior of a
generic measured variable Y is sketched with a continuous line, whereas at the
bottom its related residual r is depicted (also with a continuous line).

The variable behavior is related to a transient maneuver from the operating
condition #1 to the operating condition #2. The values simulated by the steady-state
model for the two operating conditions are expressed as $\hat{Y}_1$ and $\hat{Y}_2$ and are represented by dashed lines. The tolerance range is here assumed the same for both the
operating conditions and the related thresholds are represented by two dash-dotted
lines in the residual behavior plot. Once the transient maneuver starts, the measured
value Y diverges from the simulated one $\hat{Y}_1$, inducing the residual to cross the
threshold range at time t_I. The residual value keeps decreasing until the model
switches from $\hat{Y}_1$ to $\hat{Y}_2$, inducing a discontinuous variation in the residual behavior.
At this new operating condition, the residual is still out of the tolerance range, but
higher than the upper threshold value. This value is then crossed at time t_{II}, with the
residual again within the tolerance range. Within the time window $[t_I - t_{II}]$, the
transient maneuver is interpreted as an undesired event. Therefore, to avoid the risk
of misunderstanding, the use of a dynamic models and adaptive thresholds is
necessary (McKenzie et al. 1998).

4.4 Inferential Process

In the previous paragraphs the main issues concerning the residual generation and
detection process have been addressed. What should be bear in mind is that once a
residual overcomes a threshold level an abnormal state is detected. The detection

process shows a binary nature: if the residual is within the tolerance range, the system may be assumed in normal conditions, otherwise a faulty event could be taking place. The previous description can be translated in a mathematical formulation by introducing another feature, the *analytical symptom*, as follows (Escobet et al. 2009; Polverino et al. 2015):

$$s = \begin{cases} 0 & if \quad |r| \leq \tau \\ 1 & if \quad |r| > \tau \end{cases} \tag{4.3}$$

In Eq. (4.3) the analytical symptom s shows a binary nature: on the one hand, if the residual r remains within the tolerance range (i.e., its module $|r|$ is less or equal to the defined threshold level τ), the related symptom is zero; conversely, if the residual overcomes the threshold τ, the symptom becomes 1.

For each monitored variable, and thus for each related residual, a symptom can be associated to evaluate the state of that variable. When a symptom is active (i.e., is 1) an undesired (faulty) state may be occurring in the system. According to the previous description, any residual drift activates a symptom; then, all the information can be collected into a symptom vector. Once this vector is built, the detection process ends with the following status check: if the symptom vector has all 0, the system is working in normal condition, while, if at least one symptom is 1, an undesired behavior may be occurring in the system.

Once an abnormal state is detected, the following step to be performed consists in the identification of the faulty system component(s) (i.e., isolation process). To fulfill this task, the symptom vector can be compared to reference information linking possible faults to the collected symptoms (i.e., the monitored variables). According to Isermann (2005), one way to determine these relations consists in performing dedicated experimental campaigns, or gather significant experimental data, related to specific faults that can occur in the system under evaluation. Through this process, an explicit knowledge database concerning fault-to-symptoms association can be made. However, an a priori knowledge can be also exploited, so as to avoid complex experimental activities, which are not always feasible. Indeed, in many cases, system components are well known in the literature, and many details on their behavior and interactions can be easily found or derived by means of physical models. This approach can significantly improve the identification of the correlations among possible faults or failures and their corresponding symptoms. One possible approach that can be used to define the causal relationships among faults and symptoms is the fault tree analysis (FTA). The FTA approach can be described as a heuristic methodology correlating a fault to a set of symptoms. The main outcome of the FTA is a matrix, known as fault signature matrix (FSM), which can be used to fulfill the isolation task. The main properties and issues concerning the development and application of the FTA approach and the FSM are given in the following.

4.4.1 Fault Tree Analysis

The fault tree analysis can be defined as an analytical deductive approach, which outlines all the possible behaviors a fault or malfunction can show in the system under investigation. For a general description of the technique the work of Veseley (1981) is suggested, and applications to SOFC are reported in the papers of Arsie et al. (2010) and Polverino et al. (2015). This methodology is generally used to express how a fault can occur in a system and how it can influence the system components. A graphical representation, known as fault tree, is designed to depict these information in a clear and practical way. A fault tree consists of logical connections linking together specific events at different levels. The design process of a fault tree starts defining a specific fault (*top event*) and then investigating, through a top-down approach, all the possible causes (*intermediate events*), from which the considered fault can result. Each intermediate event is further investigated and related to other faults, until no further development can be performed. The faults at this bottom level are thus defined as *basic events*, and represent the symptoms related to the fault considered as the top event of the tree. This process is based on the physical knowledge of the system and follows a top-down approach, going from the main fault to the symptoms. It is worth noting that a fault tree does not represent all the possible system faults and their causes, but only those assessed by the analysts developing the fault tree (Veseley 1981).

A schematic representation of a generic fault tree is given in Fig. 4.6, where each aforementioned element is depicted. In this picture it is possible to appreciate the difference between the fault tree design process and its application during a diagnostic task. As previously mentioned, the former activity starts from the top level

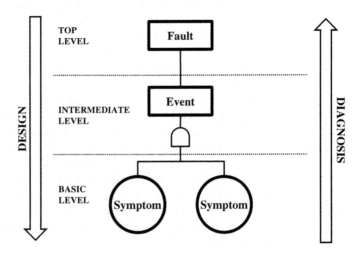

Fig. 4.6 Schematic representation of a generic fault tree: the design process follows a top-down approach—from the top level (fault) to the basic level (symptoms)—whereas the diagnosis process follows a bottom-up approach—from the symptoms to the fault

(the considered fault) and goes through the lower levels (intermediate events and symptoms), identifying each correlation leading to the basic symptoms through a top-down approach. The latter process deals with the exploitation of the fault tree for the diagnosis process, following a bottom-up approach, gathering all the symptoms and observing which fault is linked to them.

The connections among the events are expressed through Boolean operators, called *gates*, allowing or hindering to go through each level of the fault tree. A list of the symbolism generally used in the fault tree development process is given in Table 4.2 (Veseley 1981). From the table, it is possible to point out that in Fig. 4.6 an AND gate is used. This gate connects the two bottom symptoms to the upper intermediate event; it means that the specific event occurs only if both symptoms are active.

It is worth observing that, if one or more symptoms are related to different faults at the same time, proper symptom redundancy is required. This may occur either in complex or even in simple diagnostic cases. Furthermore, it must be recalled that this methodology gives only a qualitative correlation among the faults and the symptoms (Veseley 1981). For this reason, the major drawbacks of the FTA can be ascribed to the inability to: (i) detect faults that are not considered in the analysis and (ii) distinguish between incipient and severe faults (Lo et al. 2004; Polverino et al. 2015). To overcome these limitations, an exhaustive and deep knowledge of the system components and their most probable undesired behavior, in combination with improved diagnostic methodologies, is required.

The fault trees drawing process is particularly significant for choosing the variables that should be measured or estimated (e.g., not available or feasible measurement) in the system under investigation. The selection of the variables to be monitored should take into account their significance (i.e., the number and type of faults they are related to) and their measurement costs and feasibility (Polverino et al. 2015). The fault tree development process should focus on all possible faults (or at least the most relevant ones) a system may experience. This approach ensures

Table 4.2 Example of fault tree events and gates symbols (Veseley 1981)

◯	basic event/ symptom	An event not requiring further development
⬭	conditioning event	Specific rule applying to any logical gate
◇	undeveloped event	An event not further developed due to insufficient consequence or lack of information
▭	top/intermediate event	An event linked through logical gates to other events
△	transfer	Indicates a link to other fault trees (duplication avoidance)
⌂	and	The upper event occurs if all the related bottom events take place
⌂	or	The upper event occurs if at least one of the related bottom event takes place

collecting all the major variables to be monitored. Indeed, once all the fault trees are developed, a complete list of the variables (i.e., the affected symptoms) is established, gathering all the involved variables in a list where each of them appears once. The correlation among faults and symptoms defined through the fault trees can be then merged into a matrix, the so-called fault signature matrix.

4.4.2 Fault Signature Matrix

A fault signature matrix (FSM) is a two-dimensional matrix, whose rows list all the faults considered during the FTA, whereas the columns inscribe all the collected symptoms, each referring to a specific system variable. This matrix represents the final outcome of the FTA and can be directly used as reference for the isolation process. An illustrative example of FSM is given in Table 4.3 and its application procedure to fulfill a complete diagnostic procedure is presented in Fig. 4.7.

The availability of reliable information embedded into the FSM is of paramount importance to perform an effective diagnosis. Moreover, a univocal link among symptoms and faults is required: to have an unambiguous isolation of the occurring fault, each row of the FSM should be independent from one another. This means that the fault vectors must point toward different regions in the symptoms space. Therefore, an analysis on the fault vectors may provide useful insights on the robustness of the designed diagnostic strategy. Once these requirements are assessed, the first step toward the isolation of the faulty component(s) consists of

Table 4.3 Example of fault signature matrix linking each fault f to the related symptoms s

	s_1	s_2	s_3	s_4	s_5
f_1	0	0	1	0	1
f_2	1	1	0	1	0
f_3	0	1	1	0	0

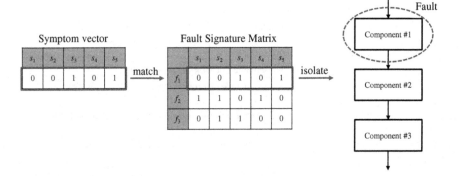

Fig. 4.7 Example of FSM application for fault isolation

comparing the symptom vector, introduced in Sect. 4.4, with the defined FSM. The vector is compared to each row of the FSM and the fault isolation is fulfilled once a perfect match occurs. Matching a row of the FSM means that the ascribed fault may be occurring in the system (with a certain probability) and the related faulty component(s) can be isolated. From the above-mentioned comment, it is clear that if two or more rows of the FSM show the same pattern, the symptom vector will not achieve a univocal matching, thus hindering the correct isolation of the fault.

It is worth recalling that an FSM developed through an FTA is mainly based on heuristic knowledge, taking into account only qualitative relations among the collected symptoms and the considered faults. Therefore, the direct implementation of such an FSM within a diagnostic algorithm may lead to a nonoptimal isolation process (Polverino et al. 2015). Indeed, according to the FTA, when a fault occurs, each symptom accounted by the fault tree should be theoretically activated. However, when dealing with a real system, not all variables are influenced by the occurring fault in the same way: some of them can clearly deviate from their nominal behavior, whereas others may show only a small change. In this case, the residuals related to these latter variables may remain within the tolerance range, thus not activating the corresponding symptoms and leading to a different symptoms vector compared to the expected one.

The FSM isolation capability can be enhanced through suitable model-based analysis (McKenzie et al. 1998). Polverino et al. (2015) report on the application of the method to the design of SOFC system diagnostics. In such an approach, a detailed mathematical model of the system under investigation can be used in order to simulate specific faulty states to evaluate the sensitivity of the system variables to the considered faults. This procedure allows identifying quantitative correlations among faults and symptoms, thus obtaining a more robust and reliable FSM. A schematic representation of the guidelines of this approach is given in Fig. 4.8.

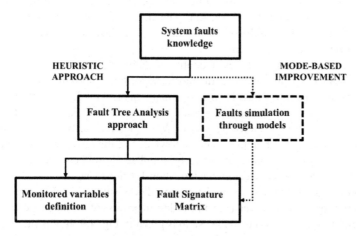

Fig. 4.8 Integration of heuristic and model-based approaches for FSM robust development, by means of FTA and faults simulation (Polverino et al. 2015)

This approach has been used in the literature, although not with a direct relation to FSM development (Simani et al. 2003). Ingimundarson et al. (2008) and Escobet et al. (2009) successfully applied that approach to PEM fuel cells. For this purpose, a system model can be run both in normal and faulty conditions to perform a parallel simulation of the system variables and thus computing the corresponding residuals. Then, these residuals are compared to specific thresholds in order to generate analytical symptoms. This process is accomplished for each considered fault. The obtained symptom vectors are then gathered all together to build a new FSM, which is thus strictly related to the fault magnitudes. Indeed, the number of the arisen symptoms directly depends on the threshold values: the lower the threshold, the larger the number (Polverino et al. 2015). Nevertheless, the same comments concerning the probability of false alarm and missed fault, proposed in Sect. 4.3, apply also for this process. Indeed, the choice of a low threshold level, although improving the isolation of incipient faults, may lead to an FSM with a large number of ones, thus increasing the risk of superimposition of the FSM row. For this reason, the FSM development should always be a reasoned application of all the aforementioned features and procedure, so as to come up with a robust and reliable diagnostic algorithm.

4.5 Case Studies

In the following subsections, two different case studies are presented, dealing with the application of the aforementioned procedures for the robust development of a reliable diagnostic algorithm based on an improved FSM and its application for real system diagnosis. Each case study refers to a specific faulty event, chosen among those of main interest concerning SOFC applications. More specifically, the system analyzed in these examples is the same sketched in Fig. 2.22 and described by the lumped model introduced in Sect. 2.3. The first considered fault copes with an increase in the mechanical losses of the air blower, whereas the second one with a fuel reformer surface corrosion.

It is worth remarking that these examples shall be taken as helpful guidelines for the development of an advanced diagnostic algorithm, through both model- and knowledge-based approaches. The use of a mathematical model can help to define a quantitative correlation among faults and variable deviation, but the use of available experimental data can be effective as well. What should be also bear in mind is that the model development requires a deep knowledge of the physics behind the faulty events under study, in order to properly simulate the effects on the monitored variables, and ensure a correct characterization for the FSM.

Before giving further details on the proposed case studies, a schematic representation of the connection between the offline design process of the diagnostic algorithm and its online application on a real system is given in Fig. 4.9. In this picture, it also represented the relations between the diagnostic algorithm and the system controller, essential for the application of suitable counteractions once a fault occurs.

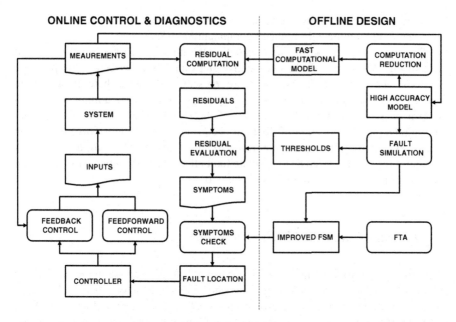

Fig. 4.9 Schematic representation of offline design of the diagnostic algorithm and its online application on a real system, with connection to system control strategies

As a first step, market requirements and industrial needs may give significant guidelines for the identification of the faults and undesired events which may affect the system during its normal operation and could also increase fuel cell degradation rate. Through this analysis, it is possible to single out the key detrimental phenomena affecting system components, and thus to define the faults to be accounted for during the fault tree analysis. It is worth noting that these information, along with the availability of experimental measurements and physical data, are also of primary importance for the choice of the mathematical model to be exploited for fault simulation and FSM improvement, as well as for the residual computation. Indeed, the accurate simulation of the physical phenomena occurring in the system (e.g., using a high accuracy model such as the lumped one described in Sect. 2.3) allows the correct identification of the variables sensitivity to the occurring faults. This phase has a twofold scope: on one hand, the definition of minimum threshold values through which it is possible to detect incipient faults; on the other hand, the evaluation of the main affected variables so as to improve fault isolation.

Therefore, as highlighted in Chap. 1, the need for real-time diagnostic tools, able to run onboard of real operating systems, as well as to detect and isolate incipient faults during system operation, pulls the current research toward the development of fast computational models (e.g., the map-based one shown in Fig. 4.1). These models retain only part of the physical information introduced in more complex models; however, by means of suitable mathematical reduction methods, it is possible to develop models whose accuracy and reliability are closer to high order ones, but with significantly reduced computational time.

The online application of the diagnostic algorithm is then presented in the left side of Fig. 4.9, where all the consecutive steps already described in the previous sections, and sketched in Fig. 4.2, are presented. What is here important to underline is the interaction between diagnostics and control aiming at optimal system operation. Indeed, the outcome of the diagnosis is the information concerning the possible location of the fault occurring in the system. This information can be exploited by the system controller to adjust both feedforward and feedback strategies (see Chap. 3), so as to mitigate fault effects on system behavior and avoid any undesired system failure (DIAMOND 2014).

4.5.1   Case Study 1: Air Blower Fault

The first fault considered in this work corresponds to an increase in the mechanical losses of the blower located at the air side (as shown in Fig. 4.10). The main purpose of the air blower is to provide the stack with the required amount of air, taken from the surroundings at ambient conditions (i.e., pressure, temperature, and humidity). The delivered air has two functions: on the one hand, it ensures the oxygen amount required for the electrochemical reaction, and, on the other hand, it can be used for cooling the stack.

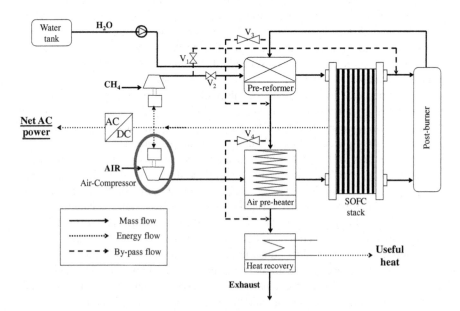

Fig. 4.10 Location of the air blower fault with respect to the system sketched in Fig. 2.22

Due to the high flow rate, the air blower can be identified as the most energy-consuming device of the entire SOFC system (Arsie et al. 2010; Polverino et al. 2015) and it is generally subject to several types of faults. For instance, its performance decrease can be due to rotating components degradation, e.g., bearings and surfaces contaminated by dirt, dust, and oils, leading to motor windings overheating and failure. Other causes could reside in motor friction increase (caused by wear), or excessive overheating (due to lack of lubricant) (Arsie et al. 2010; Polverino et al. 2015). Moreover, an increase in the system pressure drop can induce a rise in the blower absorbed power at the same flow amount. As a direct consequence, the increase in blower mechanical losses results in the growth of the absorbed electric power (which is usually provided by the stack) and of the air outlet temperature.

The application of the FTA for the description of the aforementioned fault requires the knowledge of the effects of the considered fault on the influenced component (as illustrated in the previous paragraph). Following the work of Arsie et al. (2010), recalled also in the paper (Polverino et al. 2015), the fault tree related to the air blower fault is sketched in Fig. 4.11.

Observing the proposed fault tree, it emerges how both the increase in the motor friction and the excessive overheating lead to a growth in the air outlet temperature and in the absorbed electric power (neglecting the noise amplification in the former case). On the other hand, the leakage occurring at the inlet manifold induces the decrease in the downstream pressure and in the pipe flow. From the symptoms listed at the bottom of Fig. 4.11, the variables to be monitored for the fault detection

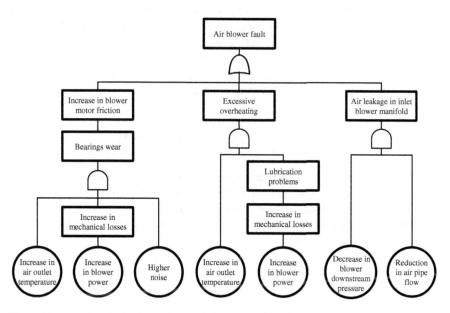

Fig. 4.11 Example of a fault tree for an air blower dedicated to an SOFC system (Arsie et al. 2010; Polverino et al. 2015)

are: (i) the temperature at the blower outlet, (ii) the absorbed electric power, (iii) the downstream pressure, and (iv) the outlet gas flow. With the definition of these variables, the description of the fault tree design procedure is completed. However, only a qualitative correlation between fault and symptoms is established. The use of the model presented in Sect. 2.3 can deepen the quantitative deviation of the observed variables once the fault occurs at a specific magnitude.

According to Polverino et al. (2015), to model the effects of an air blower mechanical losses increase, a decrease in the electric motor efficiency is introduced accordingly, as follows:

$$\eta_{cm,f} = \eta_{cm}(1 - \xi) \tag{4.4}$$

In Eq. (4.4), the term ξ is a coefficient restricted within the range [0,1] and expresses the fault magnitude: in normal operating conditions (i.e., no fault is occurring) the coefficient ξ is 0, whereas, in faulty conditions, ξ is greater than 0, with a maximum value equal to 1. This latter condition expresses the complete system failure. The term $\eta_{cm,f}$ can be introduced in Eq. (2.69) instead of η_{cm} in order to simulate the effects of the electric motor efficiency reduction on the air blower power (Polverino et al. 2015):

$$P_{cp,f} = \dot{m}\frac{c_p T_a}{\eta_{cp}\eta_{cm,f}}\left[\beta^{\frac{k-1}{k}} - 1\right] = P_{cp}\frac{1}{1 - \xi} \tag{4.5}$$

Assuming that the electric motor efficiency reduction is due to motor friction increase, part of the energy produced by the electric motor is dissipated as thermal energy. Thus, the power amount computed through Eq. (2.69) is partially used by the air blower and the residual part is dissipated. Considering that the required power is the same of that absorbed in normal conditions, the thermal loss can be computed as the difference between the blower power in faulty condition and that in normal behavior, as follows (Polverino et al. 2015):

$$Q = \dot{m}\frac{c_p T_a}{\eta_{cp}}\left[\beta^{\frac{k-1}{k}} - 1\right]\left(\frac{1}{\eta_{cm,f}} - \frac{1}{\eta_{cm}}\right) = P_{cp}\frac{\xi}{1 - \xi} \tag{4.6}$$

The thermal power computed through Eq. (4.6) is partially exchanged with the surroundings and a certain amount is transferred to the air flowing through the blower. Assuming that half of the thermal power (i.e., 50 % of Q) is exchanged with the fluid at the blower outlet, it is possible to compute the outlet temperature through the following balance:

$$\dot{m}c_p T_{cp} + \frac{Q}{2} = \dot{m}c_p T_{cp,f} \tag{4.7}$$

In Eq. (4.7), the temperature T_{cp} represents the air temperature right at the blower outlet, before any heat exchange due to faulty conditions occurs. This temperature is evaluated as follows (Polverino et al. 2015):

$$T_{cp} = T_a \left[1 + \frac{1}{\eta_{cp}} \left(\beta^{\frac{k-1}{k}} - 1\right)\right] \tag{4.8}$$

Introducing Eqs. (4.6) and (4.8) in Eq. (4.7), the air blower outlet temperature in faulty conditions $T_{cp,f}$ can be computed as (Polverino et al. 2015):

$$T_{cp,f} = T_{cp} + \frac{Q}{2\dot{m}c_p} = T_{cp}\left(1 + \frac{1}{2\eta_{cm}} \frac{\xi}{(1 - \xi)}\right) - \frac{T_a}{2\eta_{cm}}\left(\frac{\xi}{1 - \xi}\right) \tag{4.9}$$

Observing Eqs. (4.5) and (4.9), if ξ is equal to 1 the variables diverge (i.e., become infinite), meaning that a failure occurs and the system must be shut down. Moreover, the deviations of the blower power and outlet temperatures can be evaluated as shown in Eqs. (4.10) and (4.11), respectively:

$$\frac{P_{cp,f}}{P_{cp}} = \frac{1}{1 - \xi} \tag{4.10}$$

$$\frac{T_{cp,f}}{T_{cp}} = \frac{1 + (A - 1)\xi}{1 - \xi} \tag{4.11}$$

The term A in Eq. (4.11) gathers all the parameters not affected by the fault magnitude that are constant once a steady-state operating condition is reached. The behavior of Eqs. (4.10) and (4.11), function of the fault magnitude ξ, are represented in Fig. 4.12, assuming $A = 0.094$. From this figure it can be observed that in normal conditions (i.e., $\xi = 0$), the functions are both equal to 1, thus no deviation occurs. However, when the fault takes place, although the magnitude is the same, the quantitative deviation of the two variables from the normal condition is different. This example shows how each variable is differently affected by the occurrence of a fault, and highlights the need for specific threshold levels to be carefully defined for each monitored variable. It is worth noting that both the air blower power and the temperature identically deviate from the normal condition only if the term A is equal to 1. Moreover, when the fault magnitude approaches value 1, both deviations diverge (i.e., tends to an infinite value, see Fig. 4.12), thus expressing the occurrence of a failure.

To analyze the improvements achievable by introducing fault simulation for FSM development, the works of Arsie et al. (2010) and Polverino et al. (2015) can be considered. The former paper presents the development of an FSM by the only use of an FTA approach. This FSM is made of five faults and fifteen related symptoms, as shown in Table 4.4. The first fault (i.e., Air blower fault) is the same fault described in this section, referring indeed to an increase in the blower

Fig. 4.12 Representation of the influence of fault magnitude on air blower power and temperature deviation from normal conditions

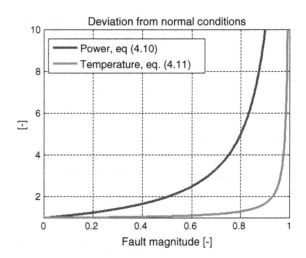

mechanical losses. It is worth observing that the symptoms pattern for this fault shows only four variables affected by the fault occurrence, that are the blower power (s_2), outlet temperature (s_8), the net power (s_3), which is the difference between the gross power from the stack and the power absorbed by the ancillaries, and the temperature at cathode inlet (s_{10}), directly affected by the temperature at the blower outlet. As commented in Sect. 4.4.2, an FSM developed through an FTA approach assumes that each symptom is active when the fault occurs, without considering the fault magnitude and the symptoms sensitivity to the fault (Polverino et al. 2015). The exploitation of a mathematical model to evaluate the quantitative deviation of the system variables, once a fault occurs at a given magnitude, can improve the FSM development. This approach has been described in Polverino et al. (2015), in which fault simulations were performed to evaluate residuals and generate symptoms at a specific fault magnitude. The symptoms have been computed by comparing the residuals to two threshold levels, one set to 1 % and the other to 5 %, in order to appreciate the changes in the FSM pattern due to fault magnitude and variable sensitivity.

Similarly to the aforementioned work, the lumped model presented in Chap. 2, improved with Eqs. (4.5) and (4.9), can be used to simulate the effect of the increase in air blower mechanical losses on the considered SOFC system. The model can be used for simulating the system in both normal and faulty conditions, emulating the scheme shown in Fig. 4.2, so as to compute percentage residuals through Eq. (4.2). In the specific case of the air blower power and outlet temperature, it is possible to analytically express the percentage residuals taking into account Eqs. (4.10) and (4.11):

$$\frac{P_{cp,f} - P_{cp}}{P_{cp}} = \frac{\xi}{1 - \xi} \tag{4.12}$$

Table 4.4 Fault signature matrix developed following only an FTA approach, adapted from the work of Arsie et al. (2010)

Faults	Symptoms	Stack power	Blower power	Net power	Stack temperature	Excess of air	Fuel temperature at anode inlet	Postburner exhaust temperature
		S_1	S_2	S_3	S_4	S_5	S_6	S_7
Air blower fault	f_1	0	1	1	0	0	0	0
Air leakage between air blower and preheater	f_2	0	1	1	0	0	0	0
Temperature controller failure	f_3	1	0	1	1	0	1	1
Pre-reformer fault	f_4	1	1	1	0	1	1	1
Stack fault	f_5	1	1	1	0	1	1	1

Faults	Symptoms	Hot fluid temperature at air preheater inlet	Air temperature at cathode inlet	Current density	Stack voltage	Air mass at cathode inlet	Temperature at anode outlet	Air temperature at cathode outlet
	Air temperature at blower outlet							
	S_8	S_9	S_{10}	S_{11}	S_{12}	S_{13}	S_{14}	S_{15}
Air blower fault	1	0	1	0	0	0	0	0
Air leakage between air blower and preheater	0	0	0	0	0	0	0	0
Temperature controller failure	0	1	1	0	1	0	1	1
Pre-reformer fault	0	1	1	0	1	1	0	0
Stack fault	0	1	1	1	1	1	0	0

$$\frac{T_{cp,f} - T_{cp}}{T_{cp}} = A\frac{\xi}{1 - \xi} \qquad (4.13)$$

Assuming a fault magnitude of 10 % (i.e., $\xi = 0.1$), Eqs. (4.12) and (4.13) show a deviation of +11.11 and +1.04 %, respectively. These values have been obtained also when performing fault simulation with the entire SOFC system model. The obtained results are shown in Fig. 4.13 with respect to the air blower power and outlet temperature. The simulation has been carried out assuming that the fault occurs at 250 s from the simulation start. From the behaviors illustrated in Fig. 4.2, it can be observed that, after the fault occurrence, the residual deviation for the blower power is about +11 % at steady state, whereas the one related to the outlet temperature is about +1 %. This confirms that the blower power has a higher sensitivity to the occurring fault with respect to that of the outlet temperature, as expected from the results shown in Fig. 4.12.

The residuals evaluation has been performed also for the other variables accounted in the FSM of Table 4.4, obtaining the deviations presented in Fig. 4.14, in which the blue bars represent the residuals. On the x-axis, the number of each monitored variable is shown, and thus its related symptom, with reference to the order followed in Table 4.4. It is worth remarking that the presented deviations have been computed once steady-state operation is reached for all the considered variables, after the fault occurrence, and only their absolute value is taken into account (their sign is here neglected). As presented in Polverino et al. (2015), the symptoms generation is strictly related to the choice of the threshold value (i.e., tolerance range). In Fig. 4.14, two threshold levels are also shown, one of about 1 % and the other of about 5 %. With respect to these threshold levels, it is possible to observe that a sensible variation occurs only for the blower power (s_2), the net electric power (s_3), and the temperature at the blower outlet (s_8), all exceeding the 1 %

Fig. 4.13 Representation of the residual deviations concerning air blower power and outlet temperature for an increase in air blower mechanical losses occurring at 250 s with a fault magnitude of 10 % (i.e., $\xi = 0.1$)

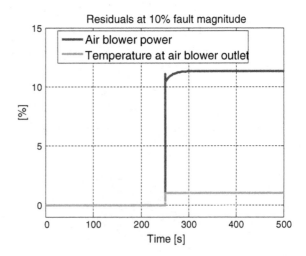

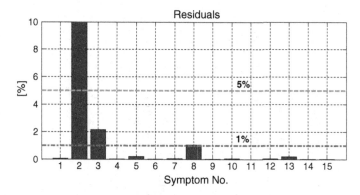

Fig. 4.14 Air blower fault simulation results: comparison of the residuals obtained for a fault magnitude of 10 % (i.e., $\xi = 0.1$) with two threshold values, adapted from Polverino et al. (2015)

threshold, whereas the other variables exhibit a really small deviation, if any. However, only the blower power overcomes the 5 % threshold level.

For this reason, the assumption of two threshold levels leads to the definition of two different symptoms vectors, as represented in Table 4.5, which are both compared to the symptom vector in the first row of the FSM of Table 4.4.

From Table 4.5 it is possible to point out that, on the one hand, the symptom vector related to a threshold level of 1 % shows only one variation, related to the air temperature at cathode inlet (s_{10}), which changes from 1 to 0. This means that this symptom is no more involved in the isolation of the specific fault. On the other hand, observing the symptom vector related to a threshold level of 5 %, it differs from the one of the starting FSM for three symptoms becoming 0: the net electric power (s_3), the temperature at the blower outlet (s_8), and the air temperature at cathode inlet (s_{10}). The observed variations are mainly motivated by the different sensitivity the involved variables show to the occurring fault, which is much more evident with a higher threshold level.

The previous results have been remarked also by the experimental validation performed in the paper of Polverino et al. (2016), in which the faults accounted in Polverino et al. (2015) have been experimentally induced on a pre-commercial SOFC μ-CHP system. In the specific case, the air blower fault has been mimicked by suitably changing the set point parameters of the system, so as to emulate the effect of an increase in blower mechanical losses. With respect to the variable listed in Tables 4.4 and 4.5, the only ones measurable on the considered systems were the stack power (s_1) and temperature (s_4) and the blower power (s_2). Their residuals, computed by comparing the current measured values with those gathered in normal conditions according to Eq. (4.2), are compared to two threshold levels, of 1 and 5 %, respectively. Such an approach allowed synthesizing the high-information content on SOFC fault behavior, as retained by the FSM shown in Table 4.5, with the fast and easily applicable on-field diagnostic method, i.e., map-based, discussed and highlighted in Sect. 4.2 and Fig. 4.1, respectively. The residual deviations and

Table 4.5 Symptoms vectors related to an air blower fault of 10 % of magnitude obtained for two threshold levels of 1 and 5 %, and comparison with the former FSM vector of Table 4.4

Air blower fault	Symptoms	Stack power	Blower power	Net power	Stack temperature	Excess of air	Fuel temperature at anode inlet	Postburner exhaust temperature	Air temperature at blower outlet
		S_1	S_2	S_3	S_4	S_5	S_6	S_7	S_8
Original FSM		0	1	1	0	0	0	0	1
1 % Threshold		0	1	1	0	0	0	0	1
5 % Threshold		0	1	0	0	0	0	0	0

Air blower fault	Symptoms	Hot fluid temperature at air preheater inlet	Air temperature at cathode inlet	Current density	Stack voltage	Air mass at cathode inlet	Temperature at anode outlet	Air temperature at cathode outlet
		S_9	S_{10}	S_{11}	S_{12}	S_{13}	S_{14}	S_{15}
Original FSM		0	1	0	0	0	0	0
1 % Threshold		0	0	0	0	0	0	0
5 % Threshold		0	0	0	0	0	0	0

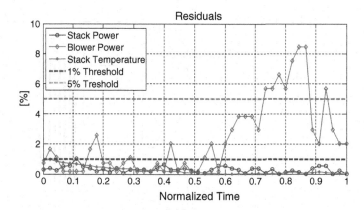

Fig. 4.15 Evaluation of absolute residuals obtained through experimental induction of the air blower fault on a pre-commercial SOFC µ-CHP system, adapted from Polverino et al. (2016)

the assumed threshold levels are illustrated in Fig. 4.15, in which on the x-axis the normalized time of the experiment is shown.

In Fig. 4.15, it is possible to appreciate a sensible deviation of the blower power once the fault occurs after a normalized time of 0.5. The other variables do not present a substantial deviation, as expected. It is worth observing that assuming a 1 % threshold level allows detecting the fault within a shorter timeframe, with respect to a threshold level of 5 %. However, the drawback of this assumption resides in the higher chance of false alarm (i.e., the blower power residuals overcoming the 1 % threshold before 0.5).

Observing the symptom vectors shown in Table 4.5, the availability of the only three aforementioned variables reduces the pattern for the specific fault to [0 1 0] for both threshold levels. The results obtained in Fig. 4.15 match exactly the proposed pattern, thus allowing a correct detection and isolation of the specific fault. However, according to Polverino et al. (2016), the reduction in the number of monitored variables can lead to the overlapping of different fault patterns, thus reducing the fault isolability. For this reason, it is worth remarking the necessity to increase as much as possible the variables redundancy.

4.5.2 Case Study 2: Fuel Pre-reformer Fault

The second case study deals with a fuel pre-reformer fault, as highlighted in Fig. 4.16. The role of the pre-reformer mainly consists in providing to the SOFC stack the required amount of hydrogen, by means of a partial reforming process applied to a hydrogen-rich fuel (e.g., methane). As described in Chap. 2, the partial

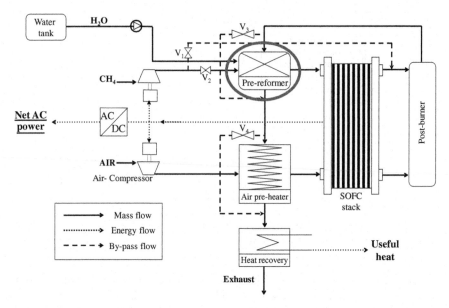

Fig. 4.16 Location of the pre-reformer fault with respect to the system sketched in Fig. 2.22

reforming exploits a specific amount of water to perform a water–gas shift reaction. Moreover, according to Fig. 4.16, the heat required to sustain the reforming process can be recovered from the postburner exhaust gases, allowing an overall increase in system efficiency.

For the proposed example, the considered fault consists in the pre-reformer heat exchange surface corrosion/erosion, which can be caused by higher operating temperatures or the presence of sulfur on active sites (Arsie et al. 2010; Polverino et al. 2015). The presence of corrosive products may foul the heat exchange surface, changing its thermal features: for instance, the adhesion of the products on the surface leads to an increase in the heat transfer resistance (Awad 2011).

As presented in the previous case study, following the work of Arsie et al. (2010), the fault tree concerning the fuel pre-reformer is sketched in Fig. 4.17, in which the left branch of the designed tree is related to the presented fault. According to the proposed representation, the main symptoms induced by the heat exchange surface corrosion/erosion consist of a decrease in the outlet fuel temperature and a change in the fuel flow outlet composition.

A representation of the heat exchange surface reduction due to fouling effects is presented in Fig. 4.18: a portion of the entire geometrical surface hinders the exchange of heat from the hot to the cold fluid. According to the simplified representation, to simulate this event, the heat exchanger surface area A_{HE} of Eq. (2.66) is reduced as follows:

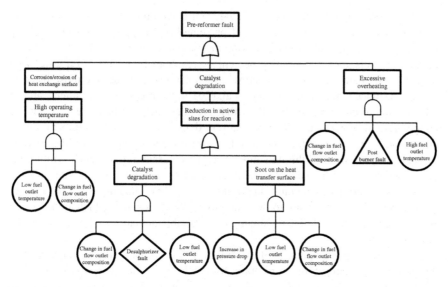

Fig. 4.17 Example of a fault tree for a pre-reformer dedicated to an SOFC system, adapted from Arsie et al. (2010)

Fig. 4.18 Schematic representation of the heat exchange surface reduction due to surface fouling: the coefficient χ represents the portion of fouled surface

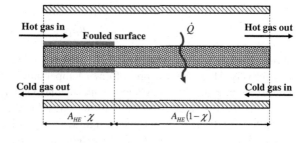

$$A_{\text{HE},f} = A_{\text{HE}}(1 - \chi) \qquad (4.14)$$

where the coefficient χ represents the portion of fouled surface and is limited within the range [0,1]. As seen for the coefficient ξ, it is related to the fault magnitude: χ equal to 0 means that the surface is completely clean and it can be entirely exploited for the heat exchange process, whereas, χ equal to 1 corresponds to a complete surface fouling, with a consequent absence of heat exchange. As a direct consequence, the reduction of the heat exchange surface affects the outlet hot and cold fluid temperatures, according to Eq. (2.66). Moreover, the induction of a variation in the pre-reformer temperature also affects the outlet fuel composition, as extensively described in Chap. 2.

It is worth highlighting that a direct evaluation of the affected variables deviations, such as those presented in Eqs. (4.10) and (4.11), cannot be performed, since the heat exchanger model here presented involves coupled nonlinear differential equations. For this reason, to suitably analyze the overall effects of this fault on the system under study, the lumped model presented in Chap. 2, improved with Eq. (4.14) can be exploited to perform system simulation in both normal and faulty conditions. This allows computing percentage residuals by means of Eq. (4.2) with respect to the main variables of interest, which can be found among those listed in Table 4.4. Indeed, according to Arsie et al. (2010), the proposed fault affects almost all the monitored variables but five, that are the stack temperature (s_4), the air temperature at blower outlet (s_8), the current density (s_{11}), and the temperatures at anode and cathode outlet (s_{14} and s_{15}). It is worth recalling that the FSM presented in Table 4.4 is developed through an FTA approach assuming that each symptom is active when the fault occurs, without considering the fault magnitude and the symptoms sensitivity to the fault (Polverino et al. 2015). The purpose of the activity here presented is, once again, to illustrate how through a suitable fault simulation it is possible to correctly define the quantitative variable deviations with respect to the fault magnitude.

Assuming a fault magnitude of 10 % (i.e., $\chi = 0.1$), the residuals percentage of the pre-reformer temperature and outlet hydrogen molar flow, expressed as:

$$\frac{T_{ref,f} - T_{re}}{T_{ref}} \tag{4.15}$$

$$\frac{\dot{n}^0_{H_2,f} - \dot{n}^0_{H_2}}{\dot{n}^0_{H_2}} \tag{4.16}$$

show a deviation of −1.25 and +0.26 % at steady state (dark blue bars in Fig. 4.19), respectively. However, the pre-reformer temperature residual results being the only one exceeding the 1 % deviation, with all the other residuals being much lower compared to this value. According to these results, it is clear that assuming a fault magnitude of 10 % does not give a significant deviation of the observed variables, almost hindering the possibility to detect the fault. For this reason, to further investigate this fault, an increased magnitude (i.e., 50 %) is chosen, so as to allow a better detectability. In this case, the pre-reformer temperature and the hydrogen molar flow residuals deviate to about −9.82 and +2.04 %, respectively (light yellow bars in Fig. 4.19).

As done in the previous case study, the residuals evaluation has been performed for all the variables in the FSM of Table 4.4, obtaining the deviations presented in Fig. 4.20. In this picture, the blue bars represent the residuals and the dash-dotted and dashed lines the 1 and 5 % thresholds, respectively. On the x-axis, the number of each monitored variable is shown, and thus its related symptom, with reference to the order followed in Table 4.4. The deviations presented in Fig. 4.20 have been computed once steady-state operation is reached for all the considered variables. With respect to the chosen threshold levels, a significant variation is shown only for the fuel temperature at anode inlet (s_6), which exceeds the 5 % threshold, with all the other residuals being much lower. The 1 % threshold is indeed reached only by

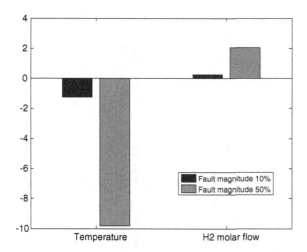

Fig. 4.19 Representation of the residual deviations concerning hydrogen molar flow and pre-reformer temperature for a pre-reformer heat exchange surface corrosion/erosion occurring with a fault magnitude of 10 % (i.e., $\chi = 0.1$—dark blue bars) and 50 % (i.e., $\chi = 0.5$—light yellow bars)

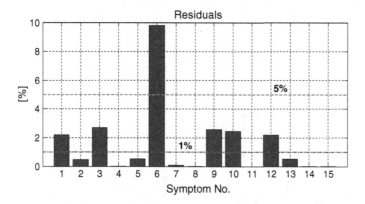

Fig. 4.20 Pre-reformer fault simulation results: comparison of the residuals obtained for a fault magnitude of 50 % (i.e., $\chi = 0.5$) with two threshold values, adapted from Polverino et al. (2015)

the stack power (s_1), the net electric power (s_3), the hot fluid temperature at the air preheater inlet (s_9), the air temperature at the cathode inlet (s_{10}), and the stack voltage (s_{12}). The remaining variables show a negligible or absent deviation.

The assumption of two threshold levels generates two different symptoms vectors, as shown in Table 4.6, which are both compared to the symptom vector in the fourth row of the FSM of Table 4.4. At a first glance it can be observed that both the new symptom vectors show differences with respect to the original one. Concerning the symptom vector for a threshold level of 1 %, four symptoms change from 1 to 0: the blower power (s_2), the excess of air (s_5), the postburner exhaust temperature (s_7), and the air mass at cathode inlet (s_{13}). However, observing the residuals

Table 4.6 Symptoms vectors related to a pre-reformer fault of 50 % of magnitude obtained for two threshold levels of 1 and 5 %, and comparison with the former FSM vector of Table 4.4

Pre-reformer fault	Symptoms	Stack power	Blower power	Net power	Stack temperature	Excess of air	Fuel temperature at anode inlet	Postburner exhaust temperature	Air temperature at blower outlet
		S_1	S_2	S_3	S_4	S_5	S_6	S_7	S_8
Original FSM		1	1	1	0	1	1	1	0
1 %Threshold		1	0	1	0	0	1	0	0
5 %Threshold		0	0	0	0	0	1	0	0

Pre-reformer fault	Symptoms	Hot fluid temperature at air preheater inlet	Air temperature at cathode inlet	Current density	Stack voltage	Air mass at cathode inlet	Temperature at anode outlet	Air temperature at cathode outlet
		S_9	S_{10}	S_{11}	S_{12}	S_{13}	S_{14}	S_{15}
Original FSM		1	1	0	1	1	0	0
1 % Threshold		1	1	0	1	0	0	0
5 % Threshold		0	0	0	0	0	0	0

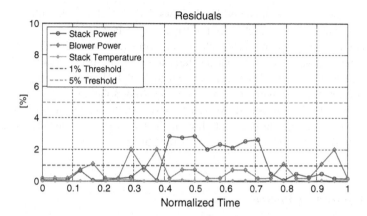

Fig. 4.21 Evaluation of absolute residuals obtained through experimental induction of the pre-reformer fault on a pre-commercial SOFC μ-CHP system, adapted from Polverino et al. (2016)

deviation in Fig. 4.20, the only significant difference resides in the postburner exhaust temperature (s_7), which seems not to be really affected by this fault, whereas all the other differences are due to the low influence of the fault. Specifically, the postburner exhaust temperature increases only by less than 1 °C, resulting in an almost zero residual. This effect can be explained considering that the stack temperature is kept near its set point, and the variation in the fuel composition has a negligible effect on the postburner inlet molar flow. The same conclusions carried out for the 1 % threshold symptoms vector can be extended also to that of a 5 % threshold, which shows five more symptoms being zeroed, due to the low sensitivity to the fault. The proposed results remark again the importance of a quantitative study of the faults effect to correctly design a diagnostic algorithm.

Also in this case study, the results obtained by means of model simulation have been achieved also by the experimental validation proposed in Polverino et al. (2016). As did for the air blower, the pre-reformer fault was mimicked by changing the system set point parameters, to emulate the effect of the heat exchange surface fouling. It is worth remarking that, with respect to the variable listed in Tables 4.4 and 4.6, the only ones measurable on the considered systems were the stack power (s_1) and temperature (s_4) and the blower power (s_2). The residuals computed through Eq. (4.2) are compared to the two threshold levels at 1 and 5 %, as shown in Fig. 4.21, in which on the x-axis the normalized time of the experiment is shown.

From this figure it is possible to observe that although a sensible change in the pre-reformer set points was achieved (although limited due to safety reasons) (Polverino et al. 2016), the influence of the proposed fault on the monitored variables is quite low. Only the stack power shows an appreciable deviation during the fault occurrence (between 0.4 and 0.7 on the normalized time scale), whereas all the other variables are not affected. Particularly, the blower power shows few points exceeding the 1 % threshold, but they are plausibly related to signal noise. For this reason, assuming a 1 % threshold level allows detecting the fault, with respect to a threshold level of 5 %, which does not trigger any alarm.

Observing the symptom vectors shown in Table 4.6, the availability of the only three aforementioned variables reduces the pattern for the specific fault to [1 0 0] for the 1 % threshold level. The results shown in Fig. 4.21 match correctly the proposed pattern, thus allowing a correct detection and isolation of the specific fault, but only with the 1 % threshold. As also proved with the model simulation activity, assuming a 5 % threshold does not allow neither the detection nor the isolation of the proposed fault, also with a significant magnitude.

4.6 Chapter Closure

The chapter here presented aimed at offering a thorough overview concerning the development and application of a diagnostic algorithm oriented toward online applications for SOFC systems. A model-based approach was followed for the design of the diagnostic structure. Several elements were identified as required components, which are a suitable mathematical model for the calculation of residuals during the monitoring phase, specific threshold levels for the generation of symptoms during the detection phase, and an FSM for the isolation of the fault and the location of the faulty component during the isolation phase.

The lumped model proposed in Chap. 2 was used to simulate the considered system in both normal and faulty conditions, to highlight the limitations of developing an FSM only through an FTA approach. Indeed, the simulation of system faults allows deriving quantitative relationships linking the monitored variables variations and the faults magnitudes, which are not taken into account in the FTA. The use of a complete system model allows also accounting for both direct and indirect correlations among faults and monitored system variables, to take into consideration the possible amplification/dumping of fault effects.

A suitable choice of the threshold levels was also highlighted, in order to evaluate the probability of missed faults and false alarms. This aspect is of paramount importance especially when applying a diagnostic algorithm online. Indeed, the continuous real-time evaluation of the system state of health, in conjunction with suitable control system, should provide robust inference, so as to avoid redundant alarms and reduce the intervention rate.

In the proposed case studies, two different system faults are illustrated: an increase in the air blower mechanical losses and the reduction in the pre-reformer heat exchange surface, due to corrosion or fouling. Residuals were computed for each monitored variable, comparing the values simulated by the model both in normal and faulty conditions. Then, the obtained residuals were compared to two percent threshold levels, one at 1 % and the other at 5 %, to investigate the sensitivity of the monitored variables to the considered faults. The analysis of experimental data acquired on a pre-commercial SOFC μ-CHP system by mimicking the considered faults validated the effectiveness of the proposed approach.

The use of a simple model, developed following a map-based approach, allowed the online deployment of the diagnostic algorithm, which was able to run in parallel with the system management and control unit. The results achieved with a limited number of monitoring variables also remarked the need for suitable measurements, so as to improve faults isolability and avoid any inference overlapping.

The obtained results highlighted that the development of an FSM following only a heuristic approach might lead to a nonoptimal fault isolation. Indeed, the knowledge of the real effects of a fault on the influenced variables become essential to meet the requirements and cope with the limitations of real system applications, such as the monitoring of only a limited number of variables, or the availability of measurement devices with low resolution. Especially in this last case, the exploitation of such kind of measurement devices drives the choice toward high threshold levels, which might hinder the univocal isolation of single faults, as shown assuming a threshold level of 5 %. On the other hand, by setting low threshold levels, such as 1 %, it is possible to avoid redundancy problems and detect incipient faults, but high resolution devices (i.e., high costs) and accurate monitoring models (i.e., high computational burdens) could be necessary. Moreover, the assumption of such low thresholds might increase false alarm occurrence.

The obtained results confirmed the robustness of the proposed approach for the application of diagnostic strategies aiming at preventing detrimental operating conditions of methane-fed SOFC units. This approach will allow a significant reduction in maintenance costs and an increase in system performance, which are of primary importance especially for a wide diffusion of the considered system in the next future.

References

Arsie I, Flauti G, Pianese C, Rizzo G, Barberio C, Flora R, Serra G, Siviero C (2000) Confidence level analysis for on-board estimation of SI engine catalytic converter efficiency. In: Proceedings of the SYSID 2000 IFAC Symposium on System Identification, pp 137–142

Arsie I, Di Filippi A, Marra D, Pianese C, Sorrentino M (2010) Fault tree analysis aimed to design and implement on-field fault detection and isolation scheme for SOFC systems. In: Proceedings of the ASME 2010 Eighth International Fuel Cell Science, Engineering and Technology Conference, FuelCell2010, 14–16 June 2010, Brooklyn, New York, USA, FuelCell2010-33344

Awad MM (2011) Fouling of heat transfer surfaces, heat transfer—theoretical analysis, experimental investigations and industrial systems. In: Aziz Belmiloudi (ed) InTech, ISBN: 978-953-307-226-5

DIAMOND (2014) Diagnosis-aided control for SOFC power systems, project funded by the European Community's Seventh Framework Programme (FP7/2007–2013) for the Fuel Cells and Hydrogen Joint Technology Initiative under grant agreement n° 245128. http://www.diamond-sofc-project.eu. Cited 30 Aug 2015

Escobet T, Feroldi D, De Lira S, Puig V, Quevedo J, Riera J, Serra M (2009) Model-based fault diagnosis in PEM fuel cell systems. J Power Sources 192:216–223

Ingimundarson A, Stefanopoulou AG, McKay D (2008) Model-based detection of hydrogen leaks in a fuel cell stack. IEEE Trans Control Syst Technol 16(5):1004–1012

Isermann R (2005) Model-based fault-detection and diagnosis—status and application. Annu Rev Control 29:71–85

Isermann R (2006) Fault-diagnosis systems—an introduction from fault detection to fault tolerance. Springer, Berlin

Kimmich F, Schwarte A, Isermann R (2005) Fault detection for modern Diesel engines using signal- and process model-based methods. Control Eng Pract 13:189–203

Lo CH, Wong YK, Rad AB (2004) Model-based fault detection in continuous dynamic systems. ISA Trans 43:459–475

McKenzie FD, Gonzales AJ, Morris R (1998) An integrated model-based approach for real-time on-line diagnosis of complex systems. Eng Appl Artif Intell 11:279–291

Montgomery DC, Runger GC, Hubele NF (2004) Engineering statistics 3rd edn. Wiley, New York

Omdahl T (1998) Reliability, availability and maintainability (RAM) dictionary. ASQC Quality Press, Milwaukee

Polverino P, Pianese C, Sorrentino M, Marra D (2015) Model-based development of a fault signature matrix to improve solid oxide fuel cell systems on-site diagnosis. J Power Sources 208:320–338

Polverino P, Esposito A, Pianese C, Ludwig B, Iwanschitz B, Mai A (2016) On-line experimental validation of a model-based diagnostic algorithm dedicated to a solid oxide fuel cell system. J Power Sources (Accepted)

Simani S, Fantuzzi C, Patton RJ (2003) Model-based fault diagnosis in dynamic systems using identification techniques. Springer-Verlag New York, Inc., Secaucus, NJ, USA

Veseley WE (1981) Fault tree handbook. Systems and Reliability Research, Office of Nuclear Regulatory Research, U.S. Nuclear Regulatory Commission, Washington, D.C

Witczak M (2003) Identification and fault detection of non-linear dynamic systems. University of Zielona Gora Press, Poland

Printed in the United States
By Bookmasters